In Came the Darkness

IN
CAME
THE DARKNESS
The Story of Blackouts

PETER Z. GROSSMAN

16548

Four Winds Press · New York

The diagram on p. 4 is adapted from The Wonder of
Electricity *by Ry Ruchlis,* Harper & Row, New York, 1965.
The diagrams on pp. 5, 34, and 35 are adapted from
Electricity 1–7—Revised 2nd edition, *edited by Harry
Mileaf, Hayden Book Co. Inc., Rochelle Park, NJ, 1978.*

LIBRARY OF CONGRESS CATALOGING IN PUBLICATION DATA

   Grossman, Peter (date)
     In came the darkness.

     SUMMARY: Presents accounts of power failures and discusses
the reasons for them and our dependency on power.
     1. Electric power failures—Juvenile literature.
[1. Electric power failures] I. Title.
TK3091.G68   621.319   81-65906
ISBN 0-590-07651-5   AACR2

Published by Four Winds Press
A division of Scholastic Inc., New York, N.Y.
Printed in the United States of America
Library of Congress Catalog Card Number: 81-65906
1 2 3 4 5 85 84 83 82 81

CONTENTS

ACKNOWLEDGMENTS

THIS BOOK COULD NOT HAVE BEEN WRITTEN WITHOUT the help of many people, organizations, and corporations. For providing the research materials necessary for the project, I'd like to thank A. M. Rosenthal of the *New York Times*, the New York Power Pool, the Mid-Continent Area Power Pool, the Electric Power Research Institute, and the Consolidated Edison Corporation. Also, I'd like to express my appreciation to the Edison Electric Institute in Washington, D.C., for its assistance. People there opened the library to me and took time to answer any questions that I posed. David Sofrin of the EEI was especially helpful.

I'd particularly like to acknowledge my gratitude to my editor, Beverly Reingold. Not only did she give considerable attention to the manuscript, but she also provided moral support and understanding from the beginning of the project.

Also, I'd like to thank Professor Edward S. Cassedy of the Polytechnic Institute of New York, who read portions of the book to check its technical accuracy and who offered helpful criticisms and suggestions.

Finally, I want to thank Polly Spiegel for reading the manuscript at various early stages and helping me find the right tone and style for the book.   P.Z.G

PREFACE

I WAS GOING HOME FROM A KARATE CLASS, WHEN suddenly the darkness fell on New York City. It happened at around 9:30 on a typical summer night. In the city, typical meant hot, humid, and unpleasant. I was lucky, though. I was riding in a car and there was a breeze coming through the window. Also, at the end of the short ride, an air-conditioned apartment awaited me. I was thinking about it when I saw the lights of the city flicker and die.

Although New York had had power failures—or *blackouts*, as we often call them—before, I had never seen the city in total darkness. Normally, it is so lit up that you can't see the stars. But on that evening, July 13, 1977, the stars came out, and buildings and streets faded into ghostly shadows. Only car headlights pierced the darkness. It was a strange and even frightening sight.

New York's streets, not very orderly on any evening, quickly became more confused than ever. There were no traffic signals, so cars slowed to a crawl. But even then it was a miracle that there weren't accidents at every corner. The sidewalks, too, were scenes of chaos. Thousands of people poured out of the then-helpless subways. They packed the sidewalks so densely that soon there was no more room on them. Many pedestrians were forced to walk in the streets.

My fellow karate student, Ray, drove slowly through the confusion but soon brought us near my apartment building. Broadway, the street I lived on, was more crowded than any place we'd seen. Ray stopped the car and looked at the people for a moment. Then he said, "I can tell you one thing: If it stays like this, the stick-up man will be out before long."

I disagreed with him. I recalled stories of a power failure in 1965. New York City had gone the entire night without electricity, but it hadn't been a bad night at all. "There will probably be less crime than ever," I replied. "In '65, people stuck together and had a good time. Things will probably go the same way tonight."

"Well, maybe," said Ray, but he didn't sound convinced. I got out of the car and said good-bye to him, and he drove off into the mess. Even if my air conditioner didn't work, I was glad to be home.

When I entered my apartment building, I was sure my friend was going to be proven wrong. My neighbors were friendlier and more cooperative than ever. Shortly after I got home, one brought me a candle. Later, another gave me dinner. The night was certain to be a replay of 1965, I thought.

But it wasn't. Ray had been a prophet, and I had been wrong. Just how wrong, I was able to see from my apartment windows. The darkness brought out swarms of "stick-up" men, women, and children. They went on a rampage of looting and destruction. Almost every store on Broadway as far as I could see was broken into and robbed. Locks and gates were seldom strong enough to keep out the determined looters, and the threat of force seemed the only protection. An all-night grocery was left untouched because the family who owned it stood guard with clubs in their hands. Stores left unguarded, however, suffered badly. By morning many of them had been cleaned out by the mob.

I watched these events with a neighbor, Dayle, and her brother, Jim. We kept asking one another, "Where are the police?" There were none around to stop the rampage. Without the police to stop it, we were afraid the violence would spread. And if it spread, where would it go? Perhaps inside the apartment buildings—inside *our* apartment building!

Suddenly six police cars emerged out of the darkness and converged on a supermarket across the street. The cars surrounded the store. Several dozen looters crawled out through the broken windows of the store and tried to get away. Some did; others were arrested. A few fought the police and were beaten to the ground. But soon the police got back into their cars and left. Only a few minutes later, the looters returned. No one bothered them again the rest of the night. But at least Dayle, Jim, and I were safe. The looters just robbed the stores; they did not enter our apartment building.

As I watched the scene on Broadway, I was amazed that this was happening simply because the electricity was out. I had never before understood how much we depend on it. Electricity does more than light our homes and run our machines. It affects what we do, how we feel, and how we behave. All of us that night were changed. Of course, the people who were looting were changed the most. To them, the end of the light meant the end of law, responsibility, and respect for other people. But the rest of us were changed, too. Most of us were afraid. Some withdrew, and some became more helpful. But we all felt and behaved differently when the electricity was gone. I realized then that we *live* by the switch on the wall.

The New York blackout of July 13, 1977, was not the first power failure or the last. Furthermore, the darkness will return. But though we know it will, we seldom think about it. We forget as soon as the power is restored and life is back to normal. We see only that the world is bright, and we live as though the switch will never fail us again.

In Came the Darkness

The Pearl Street power station, 1882   Consolidated Edison

Part I

THE ELECTRIC AGE

The Fragile Lifeline

ONE OF THE HIT MOVIES OF 1951 WAS A SCIENCE fiction thriller called *The Day the Earth Stood Still*. It was a story about a man from another planet who brought a warning to Earth. If human beings did not learn to live together in peace, he said, he would destroy every person in the world. At the urging of an Earth scientist, the alien demonstrated his power. He shut off all the electricity in the world for half an hour. Without electric power, factories came to a halt. Stores went dark and office telephones stopped ringing. Since elevators couldn't move, the upper floors of skyscrapers became almost unreachable. In homes, every electric device failed. Refrigerators warmed and irons cooled. Televisions and record players fell silent. Without electricity, the world very nearly stood still.

We will never have to face the day when the electricity of the entire world goes out for a half hour or even a minute. But the temporary loss of electric power in a neighborhood, town, city, state, or even a country is not unusual. Many of us have experienced several blackouts. Usually, when the lights in our homes suddenly flicker out, we're surprised and momentarily stare at the darkness in disbelief. Then we get out candles and flashlights and maybe we turn on a portable radio to find out how long the problem is going to last. Some electric power failures last only seconds. Others continue for hours or days. But even when they are only momentary, blackouts are annoying and disruptive. And they can be worse. So great is our dependence on electricity that even the temporary loss of it can be disastrous.

There wouldn't be a story of blackouts without this dependence. How we developed it, we shall examine in the next chapter. But first let's take a look at what we're dependent on. What is electricity?

Electricity is a form of energy that comes from forces inside all matter. Everything—the sun, the air, the oceans, even our own bodies—is made up of extremely small bits of matter called *atoms*. Atoms, in turn, are made of even smaller particles. While there are many different "subatomic" particles, two that are in every atom are the *proton* and the *electron*. An atom may have only one of each, or it may have dozens of

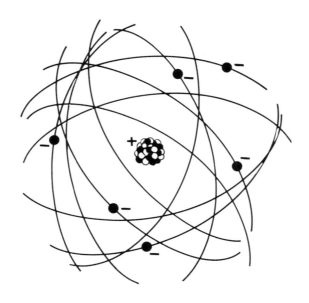

each. But every atom has both. Protons are bound in the centers of atoms and only tremendous forces can break them free. Electrons, on the other hand, move in an orbitlike fashion around the centers and are more loosely attached. What keeps them in place is a force in both protons and electrons—a force due to a property called electrical *charge*. Charge is a natural part of all protons and electrons. They possess it as long as they exist. And it is charge that makes electricity possible.

Actually there are two kinds of electrical charge. Electrons and protons have different types. The charge in the electron is called *negative* charge and the one in the proton is called *positive* charge. Lines of force radiate out from the positive proton on all sides and come in to the negative electron on all sides. The difference leads to a kind of give-and-take. The two forces complement and attract each other. So electrons are continually drawn toward the protons in the centers of atoms. And instead of flying off into space, the moving electrons usually stay in orbit.

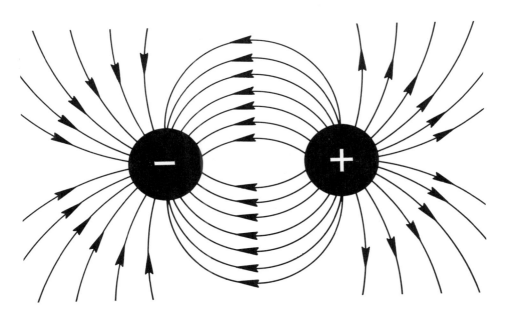

Atoms, for the most part, have an equal number of electrons and protons. When they are equal, the charges are said to be "in balance." The equal charges essentially cancel out each other and so cannot attract other electrons or atoms.

But, as we said, electrons are not firmly attached to their atoms. The attraction between proton and electron is not very strong. Electrons can be rubbed, knocked, or brushed off atoms by other atoms and by other forces. Also, new electrons can suddenly be added to atoms. But when there are either too many or too few electrons, the charges will no longer be in balance. There will be a bit of either positive or negative charge left over. This excess of charge is electricity.

While electricity is always an excess of charge, it doesn't always appear to us in the same way. It can take two different forms: *static* electricity or *current* electricity. What determines the form is the kind of material in which the excess charge occurs.

We say that matter becomes "charged" with electricity whenever its atoms are out of balance. Most things allow excess charge to build up inside them. They will lose the charge after a while. Atoms of one kind of charge will attract atoms with the opposite charge, and thus will get back into balance, sometimes very slowly. In the meantime, the charge will stay pretty much in place. And so this kind of electricity is called static electricity—static coming from a Latin word that means "to be stationary."

We can easily see how static electricity works. If we rub a balloon against a woolen sweater, the balloon will pick up electrons from the wool. It then will have too many electrons and will be negatively charged. The sweater will have too few electrons and will be positively charged. If the balloon is then put against the sweater, it will stick to it. Remember, opposite charges attract each other. But the balloon will stick for only a few moments. Electrons and protons will attract one another and get back into balance. When they do, the charges will cancel out each other, and the balloon will drop off the wool, because the force of attraction has been lost.

As we said, static electricity can build up in objects. If more and more atoms are charged and few are brought back into balance—or *discharged*, as it's called—the charge will grow. Millions of atoms have to be charged before we can see the effects of static electricity, such as the sticking balloon. As the charge adds up, the effects become more apparent. And if the charge builds sufficiently, it will move. It will leap across space in search of balance.

Sometimes an object becomes so heavily charged that the leap is dangerous. Lightning, an important example of static electricity, contains so much electric charge, so much force, that it can split trees, burn houses, and kill people. Lightning occurs when charge builds up in clouds. Billions of atoms become charged. And when the charge grows great enough, bolts of it—lightning—strike out in search of balance. They find it when they hit something. Usually what they hit is the ground. The ground is a vast reservoir of countless trillions of atoms. There are more than enough of them for the charge from the lightning to become balanced, and for its devastating power to fade.

Lightning moves fairly large distances, but it moves quickly and is soon balanced. In some materials, however, charge moves continuously. Moving electric charge is sometimes called current electricity. This form of electricity is far more important than static electricity.

Here's how it works. Say an electron is added to the end of a piece of copper wire. At once, it will enter a balanced atom in the wire. But just as particles with opposite charges attract each other, so those with similar charges push each other away. The new electron does exactly that, knocking away an electron already in place in the atom. The banished electron, however, doesn't just fly away. Rather, it moves on to another atom and it kicks an electron out of there. And that electron moves on to another atom, and the process continues all the way down the wire. With the arrival of each wandering electron, an atom becomes charged. So not only are electrons moving down the wire, charge is moving down it as well and will continue as long as electrons are being pushed into the wire.

The flow of charge is something like a current of water in a river. That's why this kind of electricity is called current

7

electricity, and the flow of charge itself is called *electric current*. When we talk about electricity, we usually are referring to electric current. Current is moving energy that can be used for dozens of tasks. It can move things, heat things, change things chemically, light things. That's why we often refer to it as *electric power*.

Electric current flows most efficiently through a narrow channel—a metal wire or cable is best. Current won't flow through all materials. Metals like copper, aluminum, silver, and gold provide the best channels for electric current. They're called *conductors* of electricity. On the other hand, current flows poorly through materials like glass, plastic, and rubber. These materials are called *insulators* and are often used to cover wires and electric equipment. Insulators help in two ways. First, because current can't flow through them, they keep current on track, flowing through the wires. Second, they act as shields. They protect our bodies and our homes from the potentially deadly power of electricity.

Yet, for all its power, electric current is uniquely vulnerable to disruption. The reason is that the current must travel in a complete circuit, an unbroken, or "closed," stream that begins at one place and eventually comes back to it. Electrons must be kept moving to produce continuous charge. If the circuit is opened for any reason, the current and the power instantly come to a halt.

The electricity in our homes originates at an electric power station sometimes miles away. Machines there start electrons jumping and bumping along. Before one light in one home can go on, the current must pass through miles of wire, over poles that support the wire, and through dozens of switches that protect the wires and the machinery. After that journey, the current enters the home, passes through wires behind walls and floors, goes into a socket—a wall outlet or a light fixture—travels on through the switch that controls the light, and only then makes its way inside the light bulb. And the trip is only half over for the current. It has to go back to the power station and the machine that started it flowing in the first place, starting the process all over again. If any part of that system fails and the current cannot complete the circuit, the light bulb won't light. There will be a blackout.

If every light bulb were a part of one circuit, anytime a bulb went out, there would be a break in the circuit and all the other bulbs would go out with it. Fortunately, though, the systems that carry electricity to our homes are designed so that doesn't have to happen. Giant circuits carry rivers of electric current to and from power stations. Branching off from the rivers are smaller streams—circuits that carry electric power to neighborhoods. Out of these streams run still smaller ones that reach each house. And within each house there are even smaller circuits, bringing electricity to every socket and switch. A break in one of these little circuits knocks out only that circuit, not the whole system. This arrangement of circuits is called *parallel* circuitry.

In Diagram A (see page 10), we can see what would happen if every light and appliance were connected to just one circuit. A break anywhere along it would knock out the entire circuit. But our electric system is arranged as in Diagram B, with many circuits. If any one of the circuits 1 through 4 breaks, only it will go out. Electricity can still go from the power station and back to it again through the other three circuits. The whole system will go out only if the circuit breaks at or near the power station. Even breaks on the outer edge of the large circuit won't knock out everything. How much goes out will depend on where the break occurs. If it happens just past circuit 1, circuits 2, 3, and 4 will go out. But power can still flow through circuit 1 and make a complete trip to the power station. If we think of each circuit as a neighborhood in a town, we can see that most people would face a blackout, but not everyone. One neighborhood would still have power.

Blackouts, then, can be very small or very large. If a circuit opens just before a light, only the light will go out. If one opens just before a house, the whole house will be without electricity, but only that house will lose it. However, if a break happens along the way from a power station to a house, the blackout may affect from two houses to a million houses. The extent of any blackout will always depend on which circuit goes out.

DIAGRAM A

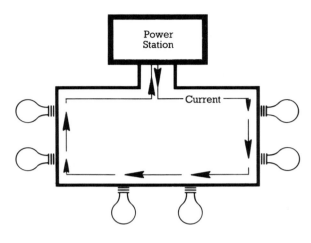

DIAGRAM B
Parallel Circuits

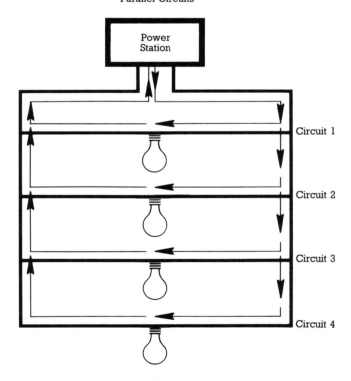

Small blackouts are very common. But large blackouts occur often, as well. Since a large blackout can create serious problems, it's frightening to think of how many things can cause one.

A simple mechanical failure can cause a disruption. Machines are never perfect. They break. Wires snap. Switches fail. Power stations stop running. Mechanical difficulties *must* occur eventually, and we should not be surprised by them.

Nature often causes blackouts, too. Hurricanes, high winds, ice, floods, tornadoes, and earthquakes can knock out stations and bring down wires and poles. While I was writing this book in upstate New York, a storm left my house without electricity for five hours. It was not at all unusual. Every year, storms bring on the darkness.

People also bring on blackouts. Human beings run machines and power stations, and humans make mistakes. They install the wrong kinds of equipment. They connect wires improperly. They accidentally throw the wrong switches. And people don't have to be electrical workers to cause a blackout; anyone can do it. Just look outside and you'll see how. Is there a power pole out there? What if a car came by and hit it, knocking it down? You'd see the pole fall and then the wires break, perhaps sending off a shower of sparks. A moment later the lights in your house would go out. Every house, business, school, factory, and farm that used electricity from that pole would be out until the wires were repaired.

What people do accidentally, they can do on purpose. People sometimes cause blackouts in acts of sabotage or vandalism. They knock down poles, cut wires, destroy power stations. Whether they do it for political reasons, for military reasons, or for no reason at all, they can leave a town completely without electric power—perhaps for weeks.

Even economic and social changes can lead to blackouts. For example, if new businesses suddenly arrive in a city, or if people move to it abruptly in great numbers, there might not be enough electricity to satisfy all of them. Of course, more machines can be built and more power can be made available. But it takes time to do this. If demand increases suddenly, there will be shortages of electricity, and the shortages might

11

cause regular blackouts. In fact, shortages have been known to cause scattered blackouts for weeks and even months.

All of this adds up to a chilling fact: Electricity is something we all depend on. It is a lifeline, but it is fragile. Because of that, it might make sense for us to say, "Let's not be so dependent on electricity anymore." After all, if we used less electric power, blackouts would threaten us less. Instead we use more all the time, and the more we use, the greater the chance that the next blackout will mean a catastrophe.

Why don't we change? One reason is that electricity is so useful; it makes our lives easier and more pleasurable. Another reason is that most of the time we don't worry about blackouts. The switch on the wall *usually* works. Years may go by from one major blackout to the next in any one city. Blackouts just don't seem to be a problem—until they happen.

Also, we have come to regard electric current as a natural part of life. Most Americans have never known life without it. We've come to think of electric power as we think of the air. At times, it seems as though there have always been electric switches on the walls of every home.

But, of course, there haven't.

2

Electric America

AS RECENTLY AS ONE HUNDRED YEARS AGO, ELECTRIC-
ity was hardly used at all, and two hundred years before that
it was almost unknown. Beginning in the 1600s, scientists
began to study this form of energy. While they learned much
about it, they were unable to make it serve many practical
purposes. In fact, until the end of the 1700s, no one had ever
produced a continuous electric current. No one even knew
that a sustained current was possible.

But in 1800 Italian scientist Allessandro Volta demonstrated that certain chemicals and metals, when put together
in a certain way, could produce a continuous stream of electricity. Chemical reactions still provide us with some of our
electricity. All *batteries* produce current through the combinations of various chemicals.

The demonstration of continuous current was a major
event in the history of electricity, but it didn't immediately
spur the widespread use of electric power. Batteries simply
didn't produce enough power to make it possible. Even today,
though batteries are important, they can't produce enough
electricity to light even a single home cheaply. Some other way

of producing current had to be found before electric power could become a part of life.

That way was discovered by English scientist Michael Faraday. In 1831, Faraday invented a device that would make supplies of electric current inexpensive and enormous. His invention was called a *generator*—also known today as a *dynamo*—and it worked by a process Faraday discovered called *electromagnetic induction*.

Before Faraday, scientists had proven that there was a connection between electricity and a mysterious force called *magnetism*. Magnetism had been known for thousands of years. People in ancient times had discovered that certain rocks possessed the power to attract other rocks. But until the nineteenth century no one even guessed that there was a connection between this force and electricity. Then scientists learned that the two were connected. They discovered that if they wrapped a piece of nonmagnetic iron in a coil of wire and then passed current through the wire, the iron would become a magnet and remain one as long as the current flowed. They didn't know why it happened. Apparently, magnetism comes from the movement of electrons. But scientists didn't develop that theory until years after the discovery of the connection between the two forces, electricity and magnetism.

Knowledge of the connection intrigued scientists of the nineteenth century, Faraday in particular. He wondered how much of a connection there was. Electricity could produce magnetism. Could magnetism then produce electricity? What if he passed a magnet over a wire? Would it start, or *generate*, an electric current? He tried it, and electricity flowed through the wire. Exactly why it happened is not known even today. But it worked. The magnet clearly started electrons pushing and shoving one another down the wire. Because it *induced* a current, Faraday called the effect electromagnetic induction.

Later, the scientist spun a magnet through the center of a coil of wire, and he found that as long as the magnet spun, it continued to induce an electric current. The device—the magnet spinning through the coil—was the first electric generator. Since then, scientists have learned that the same effect occurs if the coil spins instead of the magnet. But as long as one spins around or through the other, there will be current. If there is no spin, the current stops instantly. Modern generators are, in principle, unchanged from Faraday's own.

Faraday's generator produced a small amount of electricity. But in time engineers learned that larger magnets and larger coils produced more and more electric power. Amounts of electric power are measured in units called *watts*. Flashlight batteries produce only a couple watts of power. Electric lights in homes typically need forty to one hundred watts. By the mid-1800s, induction generators produced a few thousand watts of continuous power. A few decades later, they were spinning out a million watts and more. And generators grew steadily more powerful. Amounts of electricity coming out of them were no longer measured in watts, but rather in 1,000-watt units called *kilowatts* (1,000 watts = one kilowatt). Today electric generators can produce hundreds of thousands of kilowatts at a time to provide electricity to entire cities. Faraday's invention of the generator marked the real beginning of the Electric Age.

The first major invention that employed electricity came thirteen years after the generator. The device, invented by American Samuel F. B. Morse in 1844, was called the *telegraph*. It allowed people to communicate instantly over long distances. The telegraph was a simple device. An operator at one end of a wire opened and closed a circuit for varying lengths of time. The opening and closing produced clicks at the other end of the wire, no matter where it was. Morse suggested a system—known today as Morse code—consisting of various combinations of long and short clicks that represented letters of the alphabet. Operators could translate the clicks into language.

The telegraph was a miracle in its time. Before it existed, news traveled slowly. Messages took days to be carried from one city to the next. But the telegraph shrank the world. Suddenly people could send messages along a wire to any point, even continents away, and they would arrive within seconds. If any two places were connected by wires, people could communicate through them. And soon telegraph wires crisscrossed the country and the world.

The telegraph brought electricity into everyday use. But it didn't bring electricity into people's homes. To use the telegraph system, which reached most towns and cities by the middle of the nineteenth century, people went into a telegraph office and sent messages to offices in other places. The messages were then copied out and hand delivered like letters. Still, the telegraph made people aware of power failures. Telegraph lines broke in storms, or were cut in wars, or simply failed. As a result, newspapers lost stories, businessmen lost opportunities, and people lost contact with friends and family. With a break in a wire, the world grew larger again. Electricity was not perfect, as people who relied on it too much eventually found out.

Yet, for most people, power failures were a minor problem. Before the 1880s, life was still mostly nonelectric. People used gas, oil, and candles for light in their homes. Another American, Alexander Graham Bell, invented the telephone in 1876, but years passed before it was widely used. And as for the other electrical devices most of us take for granted today—such as toasters and TVs—they hadn't even been imagined in 1880.

Outside the home, too, little electricity was used. When people traveled, they rode horses or steam-powered trains. If they wanted entertainment, they went to a theater and watched live performers. Streets and public places were lit by gas and oil, and machinery in factories was run by steam and muscle power.

But in 1879, Thomas Alva Edison, also an American, created an invention that was destined to become a part of the daily lives of most people in the world. The invention was the *incandescent* electric light.

Edison wasn't the first person to think of using electricity to create light. Static electricity produced glowing sparks whenever a charge became strong enough. And current, scientists realized, produced heat. Anything that produced enough heat would burn or glow. So electric lighting always seemed a real possibility.

In fact, a type of electric light was created in the early years of the nineteenth century. English inventor Sir Humphry Davy built a lamp that produced a continuous electric spark, which cast a fair amount of light. Later versions of the lamp—called the *arc* lamp—were bright and reliable, and they were used on occasion for street lighting. But arc lamps were too big and too expensive to be used in homes.

Davy also demonstrated the possibility of another kind of electric light. He passed a current through a wire, and the heat in the wire produced a glow. The light was called incandescent, but it didn't glow very brightly or for very long. For the next sixty years, other scientists experimented with the idea of an incandescent electric light, but they brought it no closer to reality. The problem was to find a material that glowed brightly and kept glowing—without burning up—for many hours. Until a scientist found a material that could do both, an incandescent light had to remain nothing more than an idea.

Thomas Edison was determined to produce a successful incandescent light. He had already established himself as a leading inventor by the time he tackled the problem. He had made numerous improvements in the telegraph, and he had amazed the world with the first phonograph (nonelectric). But the incandescent light proved to be Edison's greatest challenge and his greatest achievement.

It took the inventor months of painstaking effort to create his light. He passed current through hundreds of substances. Some produced a glow but burned up quickly. Those were useless. Others glowed much longer, but the glow was too weak. He had to throw those out, as well. For the lamp to be useful, it had to glow as brightly as light cast by burning oil or gas. Edison searched on.

Finally, he found what he was looking for. He ran electricity into an airless bulb of glass. Inside the bulb he had placed a carbon thread connected to two metal wires. Because the bulb was airless, the thread couldn't burn; burning requires oxygen. But it glowed powerfully for forty hours. That airless glowing bulb was the first example of what today we call the light bulb.

Edison knew that to make use of electric light, people had to have a steady supply of electricity to keep bulbs glowing. His next step was to find a way to provide it.

Edison saw at once that he had an unusual problem. He wanted to supply electricity to every possible user of electric light. Who were they? Everyone in the world! Edison felt that businesses, factories, and a few wealthy individuals might buy their own electric generators. But certainly there was no hope of putting such expensive, elaborate equipment in every home. How was he going to get electricity to homes? He decided the answer was to build large central power stations with giant generators from which he would run wires into every home and business. Customers then would pay for the electricity they used. Simple? On the contrary. The cost of building such facilities was huge, and the technology really didn't exist. Nevertheless, Edison was convinced he could build hundreds of these stations and run wires into millions of buildings.

Edison quickly went ahead with his plans. He designed every part of the system, from the enormous generators to the meters that measured how much current people used, down to the small sockets for the light bulbs. What's more, he persuaded investors to put up the money to build a station in New York City. In 1881, construction began on the first central power plant in America. Located in downtown New York on Pearl Street, the station opened for business on September 4, 1882.

Though Edison had high hopes for the future of electricity in 1882, he was a long way from convincing other Americans of its practicality. The Pearl Street station began operations with a grand total of fifty-nine customers, and the station's service didn't immediately encourage many new ones. During the first few months, it was so unreliable that Edison and the other owners of the station decided to let customers use as much electricity as they wanted free of charge. Generator breakdowns were frequent. Switches broke and wires snapped. Even the light bulbs were unreliable. Some blew out moments after installation. Since the station also sold the bulbs in those days, station workers spent much of their time taking fresh light bulbs to customers.

Over the next ten years, business improved. But it didn't

become especially good for New York's electric company or for the electric companies that sprang up in other cities. Electricity was a luxury, and it was used almost exclusively for light. Many power stations operated only at night. There simply weren't enough electric mechanisms to justify keeping them going during the day.

In time that changed. Edison, by improving the design of light bulbs and power stations, made electricity increasingly less expensive to use. And he and other inventors began to find more and more reasons for people to bring electricity into their lives.

Starting in the late nineteenth century, men tinkered in workshops and basements, as well as inside large factories, to find new ways of using electricity. Much of their early work was designed for businesses, not for individuals. Inventors perfected electric motors and attached them to machines that previously had run on steam or muscle power. Soon looms were weaving cloth with electric power. Conveyors were carrying equipment and products with the help of electricity. Electric machines were grinding, pulling, and pushing, transforming materials in factories across America. And with these machines, American industry thrived.

Outside of factories, there were more electric marvels. Electric trains, called *street cars*, rolled through dozens of cities. In other cities, construction crews tore up streets and tunneled underneath them to make way for underground electric trains known as *subways*. Electric train transportation changed American cities and American life. People could now commute easily to work or travel considerable distances to shop. Electric trains brought every corner of a city within reach of all its residents.

By the dawn of the twentieth century, thousands of people had installed electricity in their homes. And while most of them used it only for light, electricity clearly had become a part of life. Power stations were springing up in town after town. Over 2,800 central stations were operating by 1902, and more were being built. It was difficult to find an American town where electric lights did not glow each evening.

There were power failures in those days—frequently. Nature and poor equipment knocked out circuits all the time. Lightning probably caused more blackouts than anything else did. Lightning, which you will remember is electricity, can wreak havoc with power equipment.

Let's consider what can happen if a bolt of lightning hits a wire with current flowing through it. Of course, insulation on the wire might keep the lightning from penetrating. But lightning is so powerful it can penetrate right through the insulation to the wire. Most electric power lines aren't coated to begin with, and so are completely vulnerable. And if the bolt goes through insulation or strikes where there isn't any, its electricity will flow through the wire. Conductors don't make judgments; they don't say, "I should carry this electricity but not that." They carry it all. So when a wire picks up power from a bolt, the power level in the wire rises greatly. Generators, switches, wires—all electrical equipment is designed to take only so much electricity. Thus the extra current from a lightning bolt can cause damage to every part of a circuit. And if any part is damaged it may open, resulting in a blackout. Actually, on most circuits there is protective equipment that measures amount of current, and if the level gets too high, the equipment automatically opens the circuit to prevent damage. Either way, though, the circuit opens and a blackout follows.

Power companies have learned to deflect lightning to some extent. Power lines frequently have two sets of wires on them. The first are the energized wires carrying the current. The others are called *static wires*. Static wires shield the energized wires and can capture lightning and carry it down to the ground, where it can discharge harmlessly. Since lightning strikes the highest point in an area, static wires are always above the power-carrying wires. Even with static wires, lightning can sometimes get through and hit the real power lines. But in the earliest years of electric power, lines sat totally exposed to lightning, and whenever it hit, it brought the darkness with it.

Even though power failed often then, a failure was seldom more than a minor annoyance. Since people had switched to electricity only a few years earlier, they found it easy to return to nonelectric living for a while. The only thing they lost in their homes was electric light, and they coped easily by using candles and lanterns.

Also, failures usually struck small areas. Power stations served neighborhoods, for the most part. In Chicago, for example, there were dozens of power companies operating separate neighborhood stations. When a station broke down, only one neighborhood went out. It was impossible for a failure to black out a whole city or even a large part of one.

In the 1910s, most people were still using electricity only for light. But in that decade, things began to change. Electric companies started trying to sell people on other electric devices. They had trouble at first. Few people understood electricity, and many feared it. They'd heard stories about people dying from shocks, and some thought all electric appliances and machines were dangerous, even when they were unplugged! One salesman recalled selling an electric iron to a homemaker. When he went back some time later to see how she liked it, he saw that she wasn't using it at all; she was afraid of shocks. He had to touch the plug and the other electrical connections to prove to her that the iron wasn't dangerous. Only after the homemaker saw that he didn't die was she willing to use it.

By 1920, the electric companies no longer had to prove that electricity was safe. Americans had faith, and they began to buy many electric appliances. Throughout the 1920s, they bought and bought, and by 1930, nineteen electric appliances were common in America. They included things we take for granted, like the vacuum cleaner, the washing machine, and the coffee percolator. Of the nineteen appliances, the one people wanted most was the radio.

Guglielmo Marconi had discovered the method of sending signals through the air in 1895. Within another few years, scientists had discovered that they could send words and music through those signals. And before the start of World War I (1914), wireless communication was used between ships at sea and between ships and shore bases.

But wireless communication had its greatest impact after 1920. People learned that with receiving sets—radios—

they could hear actors, musicians, and political leaders in their own homes. Radio became one of the principal forms of communication and entertainment in the 1920s. Just a decade earlier, home entertainment had meant that people sat together and played music or read aloud to one another. Now millions sat home every evening, listening to voices coming out of a box. But radios needed electricity, and there weren't any battery-powered radios in those days. Every box had to be plugged into an electric outlet.

Business, also, had grown to love and need electricity. It had helped increase factory production dramatically, and businesses, as a result, made tremendous amounts of money. With that money, they expanded and used more electricity. The more they used, the more money they made. Business grew and grew.

So did the size of buildings. With the help of the electric elevator, invented in 1887, they became taller and taller. By 1913, America had a building that soared 60 stories. Twenty years later, there were many as high or higher. Tallest of all was the Empire State Building, which was finished in 1931. The Empire State was 102 stories—1,250 feet high.

With the use of power vastly increased, power failures became more disruptive. By the late 1920s, they could stop trains, shut down factories, darken homes, and turn off radios. People had begun to feel dependent on electricity. Failures suddenly were a potential problem of real importance.

But the blackouts of the 1920s were never bad enough to cause great concern. The worst ones were the by-products of natural disasters. Floods and tornadoes blacked out areas for days. However, the loss of electricity was minor compared to the other effects of the disasters, and so it usually passed unnoticed. People had yet to face a loss of power that itself would disrupt seriously or endanger a community. But that blackout was due.

It happened on January 15, 1936. The place was New York City. The city that was the first to install an electric system was the first to be crippled by the loss of it. Less than fifty-four years after it had first turned on electric lights, New York saw the darkness come in.

•   •   •

January 15 was a dreary, rainy day. All afternoon the temperature hovered at around 40 degrees Fahrenheit. At 4:16 P.M., a power station at Locust Avenue and 132d Street in the borough of Manhattan suffered a mechanical breakdown. When the world's second largest station failed, about a third of the city and all of Westchester County to the north lost electricity. Although most of New York City escaped, the city faced a serious situation. It was the worst accidental power failure in the city's history, and, in terms of the number of people affected, the worst ever in the United States.

Hundreds of thousands of people went home that night along cold, dark streets to houses and apartments where the only lights were pale candles and flashlights. The many home appliances that Americans had become accustomed to using didn't work. People couldn't even listen to the radio to help them pass the dark hours.

Outside was no different and no better. Restaurants and movie theaters closed; they couldn't operate without electricity. On the streets, there was chaos. Because the traffic signals were out, cars hardly moved. On the subways and in the subway stations, the situation was even more serious. Sixty thousand people found themselves stranded in the dark. Had people on just one train panicked in the pitch-black tunnels, many would have died, and more would have been injured. Fortunately, people stayed calm.

But the possibility of tragedy loomed largest over area hospitals. Surgeons in the middle of delicate operations couldn't see where their scalpels were cutting. Most doctors had to finish working by candles and flashlights. In other hospital wards, problems were also serious. Nurses had to read thermometers by match light. Doctors had to examine emergency cases by candle. And yet, miraculously, no one died because of the darkness.

New York's mayor, Fiorello LaGuardia, spent the night trying to keep the city from falling into hopeless confusion. He sent the police out to untangle traffic jams and to patrol darkened neighborhoods in case the failure brought out looters

23

and vandals. (It didn't.) The night was hectic and unpleasant for officials, as well as for most of the people who lived in the blacked-out area. There were a few, however, who claimed that the evening was actually "fun." Guests at the exclusive Pierre Hotel, for example, pronounced the candlelit night "a bit of a lark."

Westchester got its power back shortly after it went out. The Bronx, another of New York City's five boroughs and the only one to go out completely, had its power restored in an hour. The subways in both Manhattan and the Bronx didn't get electricity back until 7:45 P.M., which meant that some people had waited three and a half hours to finish their ride home. Still, parts of Manhattan stayed dark far longer. In some neighborhoods of the borough—a borough in which hundreds of thousands of people lived and worked—the power was off all night.

The next day, community leaders accused the power company, New York Edison, of incompetence and neglect. One said the company had created an electric monster it could no longer control. Though company officials tried to play down the failure, a few admitted such a paralyzing blackout could well happen again.

Yet, at the time, a crippling power failure was a new and amazing event. Few expected that they'd see one quite as big again. And the power companies of America—called electric *utilities*—were busy working on a plan to make sure they wouldn't.

3

Power
Pools

ACTUALLY, IN 1936 AMERICA'S UTILITIES DIDN'T HAVE
a specific plan to prevent blackouts. Their plan was to expand
the electric system. More electricity to more people for less
money; that was what the power companies were after. In the
process of making the system bigger, the companies believed
they would make it better, too. To them, bigger *equaled* better. If the system were big enough, they were convinced, it
would be so reliable, it would be close to blackout-free.

The plan even had a name. It was called Great Power,
and the name was appropriate. Great Power demanded the
construction of huge generating stations. It required the lay-
ing of enough wire to circle the world dozens of times. It involved a work force of hundreds of thousands of people, and it
took decades to come even close to completion. But it has
served nearly every American who has lived since the 1930s,
and it is the basis of our electric power system today.

Great though it was in size, the plan was based on a simple idea, the idea of sharing power, or *pooling*. There were
three important conditions that had to be met before the plan
could go into effect. First, there had to be a huge and ever-

increasing demand for electricity; second, the right technology had to be developed; and third, competition between power companies had to stop. As we saw in the last chapter, demand for electricity grew steadily from the late 1800s on, so that was not a barrier to pooling. But the other two conditions necessary for power sharing—the right technology and the end of competition—involved many problems and issues that took decades to resolve.

How power sharing works and how and why the problems surrounding it finally were worked out, we will see shortly. To fully appreciate the development of power sharing, we should first understand an important point about electric power: Electricity can't be stored. It must be produced at the moment it's needed.

Let's consider what this means. When is electricity needed? All day long. Even when we sleep, refrigerators must have electricity, electric clocks continue to use power, and so on. If electricity has to be generated at the moment it's needed, a power company has to be able to produce it twenty-four hours a day every day of the year. And not only does the company have to provide enough electricity to meet the needs of all its customers at any given time, but it must have more than enough ready, in case customers suddenly add new appliances or more electric lights. A little extra power must be available at all times.

In any town or city, customers might need enormous amounts of power. But providing it isn't really a problem. Generators can produce hundreds of thousands of kilowatts at a time. There isn't any problem in providing it all day, either. As long as there is something to keep magnets or coils in a generator spinning, the generator will produce electricity on demand. Falling or rapidly moving water, such as a waterfall, a river, or a man-made dam, is the cheapest way to turn a generator, and there are many *hydroelectric* power plants in the world (*hydro* means water). If hydroelectric power isn't available, generating electricity may cost more, but a company can still do it day after day, hour after hour.

The alternative to falling water is a fuel-burning engine. Any engine can turn a generator, but most utilities use engines called *steam turbines*. To run one, companies have to burn fuels like oil, coal, or natural gas to boil water—huge amounts of water. The steam from the boiled water then rushes up and turns fanlike blades. The blades keep spinning as long as the water boils, and as long as they spin, they can turn the coils or the magnets of a generator. Like any machinery, steam turbines may break down. When they do, generators stop and blackouts follow. However, engineers have improved turbines enough so that they generally do work day after day, reliably and safely.

If generating large amounts of power isn't a technical problem for the power companies, it can be a waste of money. The main reason is that people's electric needs change greatly from hour to hour and season to season. But if a power company intends to generate all the electricity its customers need, it will have to have generators on hand that can produce enough power for the time of greatest use—called *peak load*. If they have too few generators or ones that are not powerful enough, there will be a blackout every day at 6:00 P.M. At that hour, customers might use twice as much electricity as they do at midnight.

But what happens to the generators at midnight? Some just gather dust. Nevertheless, several generators are necessary. Here's why. Imagine a power company that needs 350,000 kilowatts to meet peak load. It might seem that one generator that could produce 350,000 kilowatts would be a better buy than three generators that together produce 350,000 kilowatts. This is not the case, however, simply because the 350,000-kilowatt generator would have to run all day, even when much less power was needed. That would be a waste of money because the turbine for a 350,000-kilowatt generator uses a constant amount of fuel. If it gets less, it can't run. So a 350,000-kilowatt generator needs as much fuel to meet a 50,000-kilowatt demand as to meet a 350,000-kilowatt demand. When the demand is very low, then, it makes more sense to run a 50,000-kilowatt generator, which uses far less fuel than a 350,000-kilowatt generator. Utilities, therefore, have to keep on hand various generators that can produce different amounts of power to meet changes

27

in demand. Naturally, one or more will sit idly most of the day and be used only during peak load.

Actually, a utility needs to have more than enough generators to meet peak load. It has to have extra power waiting in reserve to meet unexpected demand. In case of emergency—say, a breakdown of one of the other generators or a great surge in demand—it should have a generator standing by and ready to go.

Now, a steam turbine can't start running a generator immediately if it's been off. The water has to boil first and the steam has to build up. So if a generator is to be ready, it has to be warmed up well in advance. Since the demand for power can rise suddenly, one or more ready generators, called *spinning reserve*, are a requirement for any power system.

Hydroelectric plants are exceptions. In a hydro plant, an operator has only to open a gate and let water in to turn a generator. The plant can add an extra generator in seconds.

If a company without hydro power produces all the electricity for its customers, however, there can be a lot of waste. For example, imagine a company that needs five generators to meet peak load. For one hour or maybe two—a typical period for peak load—all five would run at full speed. But what would happen during the other twenty-two hours? Perhaps three would run during an average day, and maybe by early morning only one would be necessary; the other four wouldn't run at all. During peak load, a sixth generator would spin just in case, but usually it would not be needed.

There might be a seventh generator on hand, as well. If one of the five broke down or just needed a routine overhaul, then generator number seven would become the spinning reserve during peak load, as number six took over for the one that was down. There also might be an eighth generator and even a ninth. The extra generators would allow the utility to encourage use of more electricity and eliminate the possibility of shortages.

But say, for the time being, that the power company uses only five generators. Five would produce all the income and supply all the power. Naturally, the company would have paid for the others, the ones that sit most of the time. Generators and buildings to house them are expensive. Hundreds of millions of dollars is not an unusual price for a power station; each generator alone costs millions. That's a lot of money, especially for equipment that is used seldom, if ever.

Now let's return to the idea of power sharing and see why it is so attractive. Consider a part of the country where there is a city near a rural farming community and a ski resort. Power companies may own several plants and have them in a number of different towns and cities. But let's assume that each of the three locations is served by a different power company and that each company has only one plant. If the three plants operate separately, they each have to have enough capacity to meet peak load demand. But peak load is likely to occur at different times in the three locations. In the city, peak load might come on weekday evenings between 5:00 and 7:00 P.M., when people go home from work and have dinner. In the farming area, peak load might occur in the morning, when milking machines and other farm equipment are humming. And at the ski resort, peak load might come on Saturday afternoons in the winter, when the ski lifts are operating.

At those times, each power company would have more than half its generators working. At other times, however, much of the machinery would be idle. In the case of the ski resort, power company equipment would stand idle all but a few weekends a year. Still, the equipment would have to be on hand and in good working order so that it would be ready when needed. All the companies would have more generators for emergencies and for growth. And those, too, would stand idle but would have to be maintained just in case. The high costs of buying and maintaining the equipment probably would lead to high charges to customers. And high charges would discourage people from using more electricity, which is not what most power companies want.

What would happen, though, if the three power companies connected their generating plants and power lines? They would all become part of one giant circuit, all sending electricity into the same wires. The plants would act like tributaries of a river, with each feeding into a larger stream. In a sense, this *interconnection* of electric power stations would make them one big plant, only they would be in different places. By interconnecting, by sharing, the power companies would save money.

If the plants interconnected, each one would need less reserve. They could pool their reserves and send power together wherever it was needed. If there were a new or special demand in the farm country, the rural plant would draw the extra power it needed from the city and the ski resort plants. If a new demand developed at the ski resort, the plant would draw from the city and the rural area. Also, during peak load, power would shift to wherever it was needed. So each morning, extra electricity would flow to the farms; each evening it would go to the city; and on weekends in winter, it would pass to the ski lifts. Each interconnected plant would actually need less equipment, because the three would borrow—or, as is the usual case between companies, buy—power from one another. The equipment they did have would be used more efficiently, saving money.

Besides saving companies' and customers' money, interconnections also help prevent blackouts. One plant acts as a backup for the next. If our city plant went out, the other two would keep feeding power into the lines, and city customers wouldn't be left in the dark. If the three plants connected with more plants, so much the better. The loss of one station would become less and less important. If all power plants connected, then a blackout would become almost impossible. Think of it. If one plant broke down, instantly power would flow in from hundreds of others. The loss of one shouldn't even dim a light anywhere. In this way, bigger would supposedly equal better.

Power companies can save even more money if they build plants together. After all, a single plant needs only one set of plans, one piece of land, and so on. It's much cheaper to build one large plant, even when the small ones are interconnected. In 1960, a power company estimated that it saved thirty million dollars by building one large plant instead of four small ones.

A single large plant does make customers more vulnerable to blackouts. If you have three separate plants and one goes out, only the area it serves will lose electricity. If there is one plant serving three locations, all three will lose power if the plant goes out. But the solution to that is for the plant to interconnect with large plants serving other areas. They will fill in for any breakdowns.

Obviously, interconnecting makes a great deal of sense. But in the early days of electric power, utilities didn't do it. Years went by before they were willing to consider it. Interconnecting requires cooperation among companies, and at first power companies only competed with one another.

In the late 1880s and early 1890s, companies formed and built stations where no electric power facilities had existed. And every year each of these companies tried to expand its territory, adding new customers. One company might start by serving a single street. Then it would string more wires to carry electricity to another street, and another and another. But soon it would have reached streets where other companies were trying to put in lines. Then the competition would be on. One company would offer electricity for less than a competitor. But that company would respond by cutting prices even more. Then a third company might make residents an even better offer. Even small cities soon had several companies fighting over territory. Scranton, Pennsylvania, for instance, had four separate power companies into the early 1900s.

Gradually, though, power companies realized that competition was making losers of everyone. The companies had to cut prices until they couldn't make money and couldn't pay to keep their equipment in good working order. Bad machinery led to poor service, and to frequent and prolonged blackouts, so the customers lost, too. As a result, some companies got together. Or large, rich companies took over smaller ones. Soon cities had just one power company. And these large companies began building new, large plants that they interconnected with the old. The result was improved service and lower charges.

Although interconnecting was clearly a good idea, it didn't extend much beyond city limits for many years. Large companies remained competitive with one another. Four or five weak companies found it appealing to join together to become one strong company. But when two strong companies ran into each other, neither readily gave up territory or customers. The competition stayed fierce.

Also, it was difficult for companies to get together on a large scale. They simply didn't have the technology. In the early days of electric power, plants just a few miles apart found it hard to interconnect. Electricity couldn't be sent, or *transmitted*, very far. Scientists experimented for decades with the problem of long-distance transmission of electricity, and they made progress slowly. It was sixty years after the invention of the induction generator before large amounts of electricity could be transmitted across long distances successfully.

The key to the transmission of electric power was to build up sufficient pressure to push the power along. A generator might be able to put out large amounts of electricity, but if the pressure is weak, the power won't go very far. After running through a few miles of wire, the power fades. A hundred thousand watts might be reduced to a couple thousand. After another few miles, only a couple hundred would be left. On the other hand, with a lot of pressure pushing it, a hundred thousand watts can travel hundreds of miles, and few watts are lost on the journey.

To understand why, think of electrons in a wire as water in a hose. If there's nothing pushing the water, it just dribbles out the end. However, if we put pressure on the water, the liquid jets out. And the more pressure we put on it, the farther it goes. Also, more water comes out faster. In other words, if we measure water flow in gallons per minute, we get far more gallons of water in a minute from a high-pressure hose than from a low-pressure one.

Electricity works in a similar way. Turn up the pressure and power travels over a longer distance. We measure electric pressure in *volts*, and a pressure of only a volt or two carries electricity no farther than through the tiny circuit inside a flashlight. One hundred and twenty volts is enough pressure to push current through a house. But if electricity is to be transmitted a hundred miles, it has to be pushed by tens of thousands of volts. Longer distances demand hundreds of thousands of volts. With such high voltages, though, a power company not only can send electricity a great distance; it has more electricity to distribute at the other end of the line. If a company doubles the voltage in a line, it actually is able to send five times as much power through it.

Generators provide a certain amount of initial pressure to send electricity on its way. But they don't produce enough pressure to transmit it very far. Edison's first generators could manage only 120 volts. Later improvements boosted voltage at the generator to 2,000, which could carry power through the streets of a city but not from city to city.

Then, in 1892, an American engineer named William Stanley put together a device that raised voltages tremendously. Called the *step-up transformer*, Stanley's invention took power after it came out of the generator and gave it an extra push. The transformer created what was, at that time, fantastic pressure. At the first demonstration, the device boosted electric pressure to 15,000 volts.

Stanley also built a *step-down transformer*, reversing high-voltage electricity. The second transformer made it possible to take high voltages and lower them so that electricity could be used safely in homes and businesses. In the United States, home appliances require 120 volts. With much less voltage they won't work; with much more they'll be ruined. Using both transformers, a power company can produce electricity at 2,000 volts, step up the voltage to 15,000 or higher for long-distance transmission, step it back down to 2,000 to move electricity along city streets, and then step it down to 120 for home use. Stanley's inventions were major breakthroughs in the development of electric power.

Four years after Stanley demonstrated his transformers, the first long-distance transmission on a regular basis began. In 1896, the Niagara Falls Power Company began feeding electricity on an 11,000-volt line to Buffalo, New York, twenty-two miles away. The line worked fairly well, but high-voltage lines proved especially vulnerable to lightning. In the first few years of operation, bolts knocked out the line repeatedly.

Even so, the Niagara Falls to Buffalo line proved that

long-distance transmission was practical, and within a few years other long-distance lines were set up. By 1901, power traveled 142 miles along a 60,000-volt line in California. A few years later, the first 100,000-volt line went into service in Colorado. And by 1923, southern California had a 230,000-volt line in place. It was strong enough to carry thousands of kilowatts several hundred miles with little loss of power.

While high-voltage transmission was the most important technical problem standing in the way of interconnecting, it wasn't the only one. Generators had to produce the same kind of electric current. There are two kinds: *direct current* (DC) and *alternating current* (AC). Electric current, as we saw in Chapter 1, has to travel in a circuit. But in direct current, the flow is always in one direction. DC leaves from one place on the generator and returns to another place on it. Batteries produce DC. If you look at one, you will find a (−) at one end and a (+) at the other. Electrons always flow from the (−) to the (+). The direction never changes.

With alternating current (AC), the flow changes direction. Though batteries generate only DC, imagine an AC battery. Let's say that it looks just like a DC battery and has a (+) on one end and a (−) on the other. When connected, the electrons flow out of the (−) and into the (+). But since this current is AC, an instant later the direction changes, flowing out of the (+) and into the (−). And the direction continues to change over and over and over again. When the direction changes, the (−) and (+) reverse.

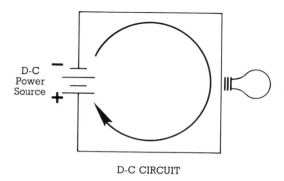

D-C Power Source

D-C CIRCUIT

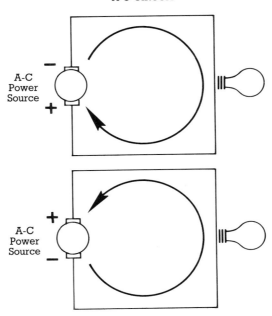

Alternating current might sound complicated, but actually generators produce it quite naturally. If a magnet and coil are set up in a certain way, every half turn of the spinning coil changes the direction of the current. A generator can be built to produce either AC or DC. But AC generators are simpler to design. Also, AC current is simpler to step up to high voltages and transmit, since transformers work only on AC. So most of the electric power in the world today is AC.

Until the early twentieth century, the advantages of AC hadn't been accepted. Edison, for one, opposed the use of AC and, for a time, he defiantly built only DC generators. However, inventor-industrialist George Westinghouse, William Stanley's employer, advocated AC generators. He eventually convinced the world of the superiority of AC. In the 1890s, nine out of every ten generators were producing DC. By 1907, however, most power companies had switched. Only one in five generators still spun out direct current.

The use of two different kinds of current delayed the growth of interconnecting. An AC generator can't hook up with a DC generator. The two types of current simply can't mix. Current traveling steadily in one direction can't flow into a line in which the current is constantly changing direction. For any two companies to interconnect, they have to agree on the kind of current they will use.

Also, if they are using AC—as nearly all do—they have to keep their generators running at the same speed. Speed is measured by the number of times the current changes direction, called *cycles*, in a second. Typically, AC changes direction sixty times—sixty cycles per second. But it doesn't have to run at sixty. It can run at sixty-two or fifty-five or fifty-eight, just about any speed. Yet before any two generators can interconnect, both have to operate at exactly the same number of cycles. If they don't, they'll be like a car with one wheel spinning at sixty miles per hour and another spinning at only fifty. The car will soon be out of control. Two connected AC generators running at different speeds will also go out of control. For a few moments, they'll "quarrel" with each other. Then, unless they stop or get back together, "in step," they'll probably both suffer damage. The difference of only a few cycles per second will cause two generators to break down.

For successful power sharing, then, generators had to be synchronized, put in step and kept there. That required control devices to keep every generator spinning constantly at the same speed. It also required a system to ensure that, in the event of a quarrel, they'd shut themselves down or break the interconnection before sustaining damage. Finally, it required rapid communication, so that one station would know right away about trouble at another. The telephone helped solve this last problem. By the twentieth century, it was a device power companies could wholly rely on.

Yet even when the technical problems of interconnecting were overcome, it didn't happen quickly. After plants in cities were connected and brought under the control of one company, power companies in neighboring towns got together and joined forces. But still the idea of power sharing had a long way to go.

Just how far became apparent in 1917–18, when America became involved in World War I. The war shocked the elec-

36

tric power industry. American businesses suddenly had to produce great quantities of material for the war. They installed new machines and expanded their factories. Most of the new machinery was electric, but there wasn't enough electricity, at least not where the factories were located. There were shortages in factory towns in many states.

At the same time, there was power to spare in places like New York City. New York's power company gladly would have sold the extra to the factory towns. But it couldn't. There weren't any transmission lines to get the power from the city to locations such as northern New Jersey, where there were shortages. The power companies realized they had to do something so that such shortages didn't happen again. The solution clearly was to interconnect.

Strong power companies who'd always competed against one another had to agree to stop competing, not to become a new, larger company, but simply to share power. To do that, companies had to agree on territory. In other words, they had to decide exactly which company would serve which area. Also, they had to agree on how power would be shared, how it would be paid for, how records would be kept, who would operate transmission lines, and so on.

After World War I, the utilities began to work out the details. The idea of one power company in any one place was accepted by everyone, customers and government leaders, as well as the companies themselves. People had decided, after years of experience, that service was cheaper and more reliable if there was no competition in electric power. Also, an interconnected system was something everyone wanted. All believed that a unified, immense electric system was essential to a growing country.

Though power companies generally became more cooperative after World War I, some conflicts remained. The most significant were those between companies owned by stockholders and those owned by government. Some city, county, state, and even federal government agencies were owners and operators of electric power facilities. But the companies owned by stockholders didn't want to have anything to do with government-owned companies. In fact, they hoped to drive the government-owned enterprises out of existence. They thought the government companies had an unfair advantage: They didn't have to pay taxes, since they were owned by the tax collector, the government. So when the government companies tried to interconnect with the stockholder utilities, the latter cried, "Unfair."

Stockholder companies also feared the government operations. If the government could provide electricity as efficiently and more cheaply than the stockholder power companies, then the government might decide to take over all electric power facilities. In other countries, the government had complete control over the electric system. Could it happen here? The stockholder companies worried.

Actually, the stockholder companies were partly to blame for the existence of the government operations. They had made electricity such an important part of life that, by the 1920s, few people were willing to be without it. But the stockholder companies had stayed in the cities and the large towns, where they could find the most customers and make the most money. They ignored huge parts of the nation, especially farm areas. If everyone in the country was going to get electric service, the government had to get into the electric power business.

Government became involved in electric power in ways besides ownership. Since each power company had a large territory to itself, the government had to make sure that none took advantage of customers. After all, there was nowhere else for customers to take their business if they felt they were being cheated by their power company. By the early 1900s, state governments had formed power authorities to see that the utilities operated fairly.

The federal government took a role in overseeing electric power after World War I. In 1920, the Congress of the United States passed the Federal Power Act. The act created a commission called the Federal Power Commission (FPC), and it gave the commission three major tasks: (1) to settle disputes between stockholder and government power companies; (2) to control transmission systems when lines crossed state borders; (3) to encourage further interconnection between companies. Congress was determined to promote interconnection on a large scale.

In the 1920s, it began to happen. New power poles and steel towers carrying high-voltage lines appeared along river beds and crossed hills and open plains. Though the stockholder/government conflict still simmered, companies began to interconnect. They formed associations with one another to develop further ties. These associations grew as companies in town after town joined. In later years, these associations formed the bases of power organizations called *power pools.*

The first pool actually began to take shape in 1926. That year, many of the utilities in Pennsylvania, New Jersey, and Maryland formed an association to deal with the problems of sharing electricity. The P-J-M Interconnection, as the pool came to be called, developed into one of the major power groups in America, serving millions of people.

Interconnections in the late 1920s brought more power to more people for less money. But what of making electric power more reliable? Was bigger equaling better? Was it helping to prevent blackouts?

The answer, it seemed, was yes. In 1927, for example, a flood on the Mississippi River knocked out power facilities in several places. But the failures were not severe and were kept localized. Thumping their chests in self-satisfaction, power company executives said that their interconnected system had done its job. Six years later, again the interconnected system came to the rescue. A storm struck Baltimore, Maryland, and knocked out its power supply. But power companies from all around sent in electricity, and Baltimore residents spent only a short time in the dark.

Then came the New York blackout of 1936, a blackout that lasted all night long. But the power companies saw no reason to blame the interconnections. In fact, power officials said that the problem was that they hadn't gone far enough. America's electric system was going to have to get *much* bigger before it became blackout-proof.

4

Grids

THROUGHOUT THE 1920S AND 30S, POWER COMPANIES interconnected and formed pools. But interconnecting didn't stop there. Pools hooked up with other pools. Together they formed huge networks of interconnected generators, transformers, and power lines—or *grids.*

These grids weren't the end of the process of interconnecting. By the late 1930s, power companies began thinking seriously about the interconnection of all grids into one grand national grid. Power experts imagined a system so vast that power from one end of America could fill in a gap left by a failure at the other end. Every generator would back up every other one. A national grid seemed the ultimate weapon against blackouts.

In the 1930s, however, power companies didn't have the technology to bring it off. Nor did they completely want to do it. The stockholder-owned power companies still resisted the idea of linking up with government-owned companies. So reluctant were they that Congress stepped in once again. In 1935, it passed another act, the Public Utilities Holding Act. A main theme of the act was to "promote and encourage" interconnections all over the country. While the legislation didn't actually force anyone to interconnect, or give the FPC power to force anyone, it left no doubt about how the government felt.

Although the FPC didn't have the authority to force interconnections, a different agency of the United States government did. And it acted in 1941. In the fall of that year, with America on the brink of war, the Office of Production Management (OPM) ordered interconnections in several southern states after power shortages developed. The order was not specifically aimed at improving the electric system. The OPM simply wanted to insure that vital defense industries in the region had enough power. But the order did further expand the interconnected power system of the region.

About one month after the OPM ordered interconnections, the Japanese attacked Pearl Harbor and America entered World War II. At this point, any power shortages posed a real danger to the country. Electricity was more vital in the Second World War than it had been in the First. By the 1940s, war required all kinds of complicated equipment, such as radar and rockets. To build that equipment, industry had to have plenty of power. And to insure that it did, the new War Production Board made it clear that it, too, would order interconnections if shortages developed.

There were not, however, any serious power problems during the war. There were blackouts, but they were of a sort not known in the United States before 1941. These blackouts were not accidental; they were *ordered* by the United States military. People had to turn off all lights, or at least cover them, so that enemy bombers in the air would have trouble seeing targets on the ground. These exercises (America was never actually bombed) were called *blackouts.* Though blackouts are of a different sort today, the word has stuck. Power companies, though, prefer less dramatic descriptions of power failures. They call them *outages* or power *interruptions.*

During the war years, the country needed power for running factories. Individuals had to use it sparingly. They couldn't even buy home appliances like washing machines and refrigerators. Factories that made them were producing guns and ammunition, instead. But in 1945, with the fighting ended, factories returned to making household appliances. And people bought. How they bought! Stores were filled with washers, phonographs, refrigerators, and a wondrous device invented in the 1920s—television. As fast as the goods came in, they went out into the hands of eager consumers. Americans craved electrical wonders, and manufacturers tried to create them. They even built showplace all-electric homes. These homes, Americans were told, were the homes of the future. All-electric became the dream of many, the dream that meant comfort and a good life.

In some parts of the country, the demand for more electricity to run these appliances grew so fast that the power companies couldn't meet it. In California and the Pacific Northwest, shortages left communities on the verge of blackouts for months. But the electric companies, encouraged by the government, built more stations and made more interconnections. Soon the shortages passed. The electric system had become big enough to meet the demand and then some. The system grew.

As it grew, Americans became more confident in it. People thought it a symbol of America's greatness. The huge electric system was evidence of our position as the strongest, richest nation on Earth. Other countries had endless problems with electricity, but not America. Our system was the best, and it was growing bigger and better.

Each year, power companies poured thousands more kilowatts into the lines. Each year, they strung new cables and made new interconnections. Though demand for power grew every year, in the 1950s it didn't outstrip the supply. Electric power was available, and it was cheap. People saw no reason to believe that they wouldn't have more in the future. Certainly, Americans never considered using less. Armed with American know-how, the electric industry would always see to it that there was enough. Americans took electricity for granted.

The extent of the interconnections staggered the mind. By 1955, power companies had stretched more than one hundred thousand miles of 110,000-volt lines over thousands of poles and towers. What's more, they'd set up thousands of miles of 230,000-volt lines, and they'd put in fifteen hundred miles of new 345,000-volt lines. These lines were powerful enough to make a national grid a possibility.

By the late 1950s and early 1960s, most utilities in America had interconnected with a power grid, and most belonged to a power pool. Some even belonged to more than one pool. Though all pools involved interconnection, they weren't all run the same way. Some were tightly organized and controlled. Companies in them worked together on all aspects of power distribution and management. By the late 1950s, there were seventeen such pools in America. These pools often joined with other pools or with independent neighboring utilities to form other, looser pooling organizations. These looser groups acted more as planning councils to establish and maintain interconnections. In some cases, the informal pools grew very large, linking the power companies of whole regions of the country.

Whether companies decided to organize tight-knit or informal pools, still they pooled extensively. Cooperation was the general rule in the electric power industry. Disputes between stockholder and government power companies, though not ended, were put aside, and pools included both kinds of companies. In 1957, for instance, forty-four utilities in the central United States met to form a pool. Of the forty-four, twelve were owned by stockholders and the other thirty-two by cities, towns, counties, or federal government agencies.

Also, at this time regional pools linked to form vast grids, and the grids linked to form still larger networks. In the 1960s, a total of 120 power companies from pools in the Northeast, Southeast, Midwest, Middle Atlantic, and Central states connected to form a massive grid. It stretched from the Rocky Mountains in the west to the Atlantic Ocean in the east, and from Canada in the north to the Gulf of Mexico in the south. Power flowed around the country along an ever-growing network of high-voltage transmission lines. Total mileage of high-voltage lines reached 350,000 in the 1960s. (That is equal to the distance from Earth to the moon and halfway back to Earth again.) Only the Rocky Mountains prevented power companies from uniting in one national grid.

The solution to blackouts was close to complete. There would always be local problems, as winds and rains knocked down power lines; there was no cure for that. But a major power failure seemed inconceivable. Power would rush in from all over the country to make up for any breakdown. The grids seemed to have made a large blackout impossible.

But soon it became all too clear that they hadn't.

5

The Cascade Effect

IN THE EARLY AFTERNOON OF JUNE 25, 1962, A
serious flaw in interconnected power systems appeared for the
first time. It did so in a spectacular fashion. At 1:30, a massive power failure struck the central United States. Parts of
eight states from Montana to Kansas lost electricity at almost
the same moment. The failure covered three hundred thousand square miles, an area the size of Italy and France com-
bined. It left three million people without electric power. Some
had to do without it for four hours.

How could it have happened? Wasn't pooling supposed to
prevent blackouts? Wasn't every system supposed to step in
and back up every other one? How could so many people lose
electricity?

The answer proved as astounding as the event itself. In-
credibly, the *solution* to blackouts had been the cause. The
grid, the system that was supposed to prevent power failures,
had itself led to a huge one. What's more, the reason wasn't
that the system had failed. Rather, the blackout became so
widespread because the system had worked perfectly! Sound
impossible? Here's what happened.

On the afternoon of June 25, a small instrument called a *voltage regulator* failed at a power station in Omaha, Nebraska. The device was designed to keep voltages at steady, precise levels, and when it failed, there was a risk that voltages would rise too high or fall too low. If either happened, generators, transformers, and even wires would suffer damage, so the breakdown of the regulator was a serious matter. When it went out, everything in the plant shut down; the equipment was turned off automatically. The shutdown took 100,000 kilowatts of electric power out of the grid. It was a substantial loss, but generators in nearby plants quickly filled in, and there was no blackout.

So far everything was okay. But a little while later, switches failed and knocked out a 230,000-volt transmission line coming out from the hydroelectric station at the Fort Randall Dam in South Dakota. Instantly, 200,000 more kilowatts left the lines. After that, the interconnected system spread the failure over 300,000 square miles within minutes.

Why? Consider how an interconnected system operates. Every plant and every generator in each plant is connected to every other one. If any one generator breaks down, the others fill in automatically. All available power, including spinning reserves, goes quickly to make up the loss. If it isn't enough, generators try to push a little more power into the lines. But generators, as we saw in Chapter 3, always operate near capacity. Although they can provide a small amount of extra power, they can't push too hard.

What if a lot of extra power is needed? Generators then try to produce still more electricity. They push harder and harder to meet demand throughout the system. But if the demand or load becomes too great, they may be forced beyond their capacity, becoming *overloaded.* Overload can damage generators severely, so generators usually are equipped with automatic protective switches. The switches turn off the generators when overload approaches. And since electricity travels very fast—at almost 186,000 miles per second—generators can reach overload quickly. Before an operator can figure out what's wrong, every generator in a plant, or even in a pool, might have shut itself down.

Now let's go back to the events of June 25, 1962. First, the power station in Omaha shut down automatically, or, as power company workers say, "tripped out." But in order to keep its customers from losing power, the company that owned the station immediately drew power from other stations on the grid, forcing generators throughout the system to push a little harder.

But then another large part of the grid—the Fort Randall Dam station—went out. There weren't enough reserves left now, although, once again, generators in plant after plant tried to fill in. But this time they couldn't produce any extra power. One by one, the generators on the grid reached the point of overload. When they did, they began to protect themselves by turning themselves off. As each station lost its generators, however, it, too, began to pull power from the grid. All plants did that automatically; that was how the interconnected system worked.

Meanwhile, the demand on the generators that were still running steadily increased. Of course, the demand was far too much for these generators to meet. Instead of filling in, they were all sucked into the failure. Soon all generators still connected to the grid in that area went out. The failure fell over them one by one, like water cascading downhill over rocks. Experts called it the *cascade effect.*

After the failure spread, it was too late for neighboring power pools to save the situation. The loss of a whole region of the country was far too much for any other pool to make up. In fact, the other pools had to cut their connections to the Central state grid quickly. Had they not, they, too, would have been pulled into the cascade, and the blackout might have spread all the way to the Atlantic Ocean.

The Central state system—then a loosely organized grid system made up of several small power pools—might never have experienced the cascade if each company had had a large spinning reserve to bring quickly to the rescue. But one of the reasons utilities wanted to interconnect in the first place was to cut down on spinning reserves to save money. Thus there was a good reason for companies not to have large spinning reserves. On June 25, 1962, however, it became clear that there was also a risk: The risk was the cascade effect.

Once it had started, there was only one way for companies on the Central state grid to escape the cascade. Companies had switches to cut themselves free of the grid without shutting down their own generators or their connections to their local customers. Had utilities thrown those switches, there would have been only small blackouts in a few places. In places where companies were producing enough power to meet local needs, there would have been no failures at all. But most operators didn't have time to throw the switches. By the time they realized what was happening, their generators had shut themselves down, and the cascade was on its way.

When it had finished, two-thirds of the state of Nebraska and most of Iowa, South Dakota, Kansas, and Wisconsin had lost power. Small parts of three other states had been hit, as well. It was the largest accidental power failure in American history to date.

The failure struck on a hot day, so everyone had to suffer through several hours without air conditioning. There were also some problems with traffic when stoplights blinked out in dozens of cities and towns. Otherwise, it wasn't a very painful experience. Perhaps that explains why the event was practically ignored by the rest of the country. Newspapers outside the affected area barely covered the blackout. The government didn't make much of it, either, and pools around the country made few changes in the way they operated. Americans certainly didn't ask if it could happen again.

For the next three and a half years, power grids operated smoothly, and the cascade effect was virtually forgotten. Officials in the electric industry and of the Federal Power Commission still regarded the United States power network as close to perfect. In 1964, FPC officials told Congress the grids were so effective that even a nuclear attack wouldn't knock one out. American technology was still in control.

Yet the time was fast approaching when America's faith in its electric systems would be shaken. The darkness was about to fall, and this time everyone in the country would know it.

Manhattan skyline during 1965 blackout and after
Wide World Photos

Part II

THE
GREAT
ONE

6

Tuesday, November 9,1965

OF ALL THE REGIONAL POWER POOLS, NONE SERVED a more vital area of North America than the Canada-U.S. Eastern Interconnection. Called CANUSE for short, the pool brought electric power to the people of the northeastern corner of the United States and parts of the provinces of Ontario and Quebec in Canada. The region was a major center of business, culture, and communication. It was one of the world's financial centers, and a major port for both ships and planes. It was the home of leading universities, famous museums, great libraries, historic landmarks, and internationally renowned performance centers. It was the headquarters of all three national television networks and most of America's magazine and book publishers. And it was the home base of many of the largest corporations in the world, including Exxon, Eastman Kodak, General Electric, Xerox, IBM, RCA, and Polaroid. Altogether, CANUSE linked more than thirty million people and eighty thousand square miles into one electric system.

Twenty-three companies belonged to the pool, and dozens of generating stations were tied together. Much of the power that flowed into the pool came from Niagara Falls and the Niagara River. The giant falls and the churning river produced several million kilowatts of cheap, reliable hydroelectric power.

Transmission lines from the Niagara power stations stretched in all directions. Some went west to Canada's second largest city, Toronto, and then on to the towns and farms farther west in Ontario. Other lines ran north along the St. Lawrence River, feeding power to rural farming communities in upper New York State, as well as Ontario and Quebec. The main lines ran east and south. A 345,000-volt transmission line crossed New York State through the cities of Buffalo, Rochester, Syracuse, and Albany. At Albany, it split into two parts. One line ran south through the vast, sprawling suburbs of the Hudson River Valley and on to the more than eight million people of New York City and nearby Long Island. The second line continued east through the towns and villages of Massachusetts, all the way to the city of Boston. Out of the main streams ran smaller lines tying to the grid the fishermen of New England, the dairy farmers of central New York, the factory workers in dozens of towns and cities, the ski slope operators of Vermont and New Hampshire, and the resort workers of Cape Cod. Lines also ran from CANUSE into the P-J-M pool. Together, the two great pools formed the northeastern power grid. It, in turn, was loosely connected with most of the power companies east of the Rockies.

CANUSE had served northeasterners faithfully for many years. The switch had seldom failed, and when it had the problems had been local. The electric system gave people a feeling of security. It was one thing they could trust.

While Americans trusted technology, they were growing increasingly suspicious and afraid of one another. On November 9, 1965, the country was not at peace with itself. Throughout the year, America had boiled over with trouble. During the summer, black people had rioted in the ghettos. Each month seemed to have brought with it new outbreaks of violence. Some radical black leaders were calling on black people to arm themselves, and many Americans feared that soon there would be war between blacks and whites. Certainly, no one believed the rioting was over. Poverty and discrimination, the seeds of ghetto violence, persisted. The question was, what next? How bad would the next riots be? Where and when would they strike?

In 1965, the American people were also troubled by a distant place called Vietnam. People had just become familiar with the name, and now they heard it constantly. Every day more Americans seemed to be touched by it. Young men from every part of the country went off to fight in a war there. By 1965, many had returned injured. Many had not returned.

While some men fought on the battlefields to win the war, others fought at home to stop it. Protest stirred across the nation. On the morning of Tuesday, November 9, one northeastern newspaper showed a picture of young men burning their draft cards. For that, they risked a jail term. On that same day, another young man protested in a way that shocked the country. In New York City, in front of the United Nations building, he set himself on fire. Later, he died.

The protest movement didn't have the support of the majority of Americans in 1965. Many people viewed the protestors as pro-Communist and anti-American. This was still the era of the Cold War, the nonviolent struggle for power and prestige in the world between the Western democracies and Communist nations. Since the other side in the Vietnam War was Communist, some Americans regarded antiwar protestors as Communist sympathizers and traitors to the United States.

So America was divided in 1965. Black people were at odds with white people. War protestors were at odds with war supporters. Americans feared and opposed one another, and, as the events of November 9 unfolded, the nation's divisions would cross everyone's mind. Many worried, at least for a moment, that something terrible was at hand.

But those thoughts came late in the day. Tuesday, November 9, started out unremarkably. It was a typical autumn day. The air was cool and the sky clear. In New York City, the temperature was around 40 degrees Fahrenheit. No one expected anything unusual. Public officials certainly didn't. Many of them were away from the region. New York State's governor, Nelson Rockefeller, was attending a conference in Atlanta, Georgia. Rhode Island's chief of state was in flight somewhere over the Pacific Ocean, and Vermont's governor was on vacation in Europe.

There was nothing unusual going on along the power grid, either. Electricity flowed normally through the Northeast. In the morning, farmers had plenty of power to do their work. Commuters trudged into the cities with no special problems. For everyone, it was an ordinary weekday morning.

And so the day went. CANUSE kept pumping electric power faithfully to the more than thirty million people of the region. There seemed absolutely no reason to worry about it. The weather wasn't the kind that strained the system. It wasn't too hot or too cold, so there was no reason for people to overuse air conditioners or heaters.

When the sun set at 4:45 that afternoon, CANUSE was ready for evening. Lights blazed on throughout the area. Workers rushed for trains and subways. People started getting dinner ready.

In Conway, New Hampshire, eleven-year-old Jay Hounsell was on his way home. He carried a stick in his hand, and as he walked along he swung it, hitting trees and power poles. At around 5:30, he struck a power pole and saw the street lights suddenly flicker out. He watched as house lights went out, as well. Soon the darkness masked everything as far as he could see. Jay was terrified. Thinking he had caused the blackout, he ran home and confessed to his mother. But she reassured him. As she said later, she knew he couldn't have put out the "whole gizmo."

Of course he hadn't. Jay had only been a witness. Like thirty million other people, Jay Hounsell would forever remember where he was on Black Tuesday when the lights went out, for the greatest power failure in history had begun.

7

The Darkness Comes

THE TROUBLE STARTED AT AROUND A QUARTER AFTER five. At the time, no one knew why or where it began. Not even the power companies knew. All they could say was that it started somewhere near Niagara Falls, for on both the American and Canadian sides of the waterfall, the entire electric system suddenly went haywire. Strangely, it seemed to begin not with a loss of power but rather with an abrupt surge of it. In stations throughout the Niagara region, power levels rose tremendously in an instant. At the Robert Moses Niagara Power Station on the American side of the falls, needles that indicated power levels practically jumped off their dials. The twelve huge generators inside the plant promptly went crazy. Some slowed down because of the surge. But, as they did, others speeded up. They went out of step and quarreled hopelessly with one another.

Every other station in the area experienced trouble, as well. Many generators shut themselves down to prevent damage. Sensing a major problem, the main line that crossed from Canada to the United States at Messina, New York, automatically opened and separated the power systems of the two countries. In only a matter of seconds, the entire power network around Niagara Falls was out of commission.

The Niagara plants were generating several million kilowatts at the time. When they went out, the whole pool was endangered. There was no way other generators along the line could make up for a loss of that size.

Had the companies of CANUSE known what was going on, they might have cut themselves free of the grid. If they had, a number of towns and cities would have escaped a blackout entirely. Without the support of the grid, a few probably would have experienced failures, but these would have been small and of short duration. The problem was that the companies didn't know what was going on. They just saw that *something* was wrong. They saw dials swing one way and then the other. But why did they jump like that? Where was the problem? Was it their own equipment? A nearby plant? The grid? Power plant operators searched frantically for answers.

But they didn't cut away from the grid. They were instructed not to do that, except as a last resort. Some companies had automatic switches to cut themselves free, but so total was their commitment to the grid that they shut off the automatic switches and stayed on the grid. And it made sense. Remember, the grid was there so that everyone could back up everyone else. To run away from the grid in a time of trouble seemed foolish—like throwing away a lifejacket as a ship is sinking.

The cascade began. Generators from Cape Cod to Canada pushed as much power as they could into the lines to make up for the millions of kilowatts that had been lost. But the demand was too great. And as technicians tried desperately to figure out what to do, the machines acted on their own. One after another they stopped to protect themselves from damage. All across the Northeast, the darkness began to fall.

It took about half an hour. Toronto and eastern Ontario went first. Shortly after the Niagara power network collapsed, that part of Canada was shrouded by the night. Then the darkness moved quickly east. Within a minute or two, it had covered Rochester, New York, and the towns near it. Moments after most of central New York State was enveloped.

Two minutes later, at 5:19 P.M., an operator of a plant belonging to the Orange and Rockland Power Company in New York made the decision to cut free of the grid. He was the only one who acted, and he saved parts of the two suburban New York counties from darkness. But not all of Orange and Rockland made it. At almost the same time that the operator pulled the switch, the rest of the area was blanketed by night.

The darkness rolled east. It covered New England next. Massachusetts and Rhode Island went under at 5:21. A minute later, more of northern, eastern, and western New York was engulfed after a six-minute struggle against the night. Now darkness covered more than a dozen major cities in the Northeast—though not the biggest. New York City still had electricity, but the wave of black was pushing down the Hudson River and the giant city was not going to escape.

At the time the trouble started, New York's power company, by then called Consolidated Edison, was generating 4.5 million kilowatts. It was also buying another 350,000 kilowatts from other companies on the grid. Had Con Edison tried to make it on its own, it might have had just enough power to keep the city from slipping into the darkness. But it did not, and as the crisis grew, power companies north of the city began automatically to pull away what electricity the city had. In minutes, New York City reached the brink of a blackout. Con Edison operators had to cut away from CANUSE at once or fall into the darkness. Like most others, however, Con Ed operators stayed on the grid, trying to figure out what was wrong and not acting until it was too late. As technicians watched helplessly, New York City's electricity faded away. At 5:27, most of the city was in darkness.

Even then it didn't stop. The blackout extended to Long Island and Connecticut at 5:30. Ten minutes later, it claimed Vermont and New Hampshire. Only then did it finally end. Switches did keep Maine from being overcome by the night, and others separated CANUSE from the P-J-M pool in time, which was extremely fortunate. Otherwise the blackout would have been much greater. It could have covered most of the nation east of the Rocky Mountains.

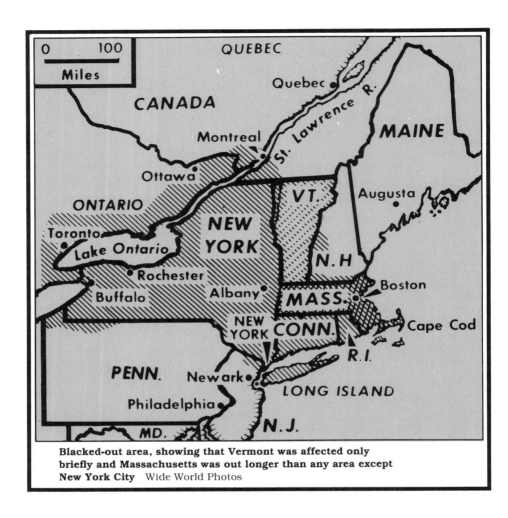

Blacked-out area, showing that Vermont was affected only
briefly and Massachusetts was out longer than any area except
New York City   Wide World Photos

Except for Maine and one section of Orange and Rockland counties, only a handful of communities in the Northeast avoided the darkness. Little Nantucket Island off the coast of Cape Cod had its own independent power company and was untouched by the blackout. Several other towns sidestepped the failure because they, too, had never hooked into the grid. One part of New York City, Staten Island, also made it through when, for unknown reasons, it separated from the rest of the city. A few towns and cities were out for only a couple of minutes, as local power companies managed quickly to regain control.

Still, the blackout was vast. It covered most or all of five states and parts of three others, as well as two Canadian provinces. Altogether an area the size of England went dark. But the most extraordinary aspect of the blackout was not the area it covered; the '62 failure had actually covered a larger one. Rather, it was the number of people who were affected—thirty million!—ten times the number who had experienced the blackout in 1962.

Few people realized how vast it was at first. Most people had experienced local power failures before, and they naturally assumed that this was another one. However, when they turned on their portable radios, they began to hear reports they could hardly believe. Radio announcers said that power was out from Maine to Florida, from Boston to Chicago. As bad as the blackout was, the reports made it seem worse. Even after the reports were corrected, people were astonished. Most had never heard of or imagined a power failure so great.

As a result, people were scared. The possibility of sabotage crossed almost everyone's mind. Some thought of war protestors. If one could set himself afire, might not others choose to destroy the electric system? Some people feared that Russian or Chinese Communists were behind it. On a commuter train outside of New York City, a conductor told passengers, "Some Commie's pulled the switch from here to Canada." And on a New York subway train, a passenger muttered, "The Russians are coming." People laughed at the idea, but mostly they were laughing away their fears.

Few totally dismissed the possibility that the blackout

was caused by sabotage. Some even worried that this was the first step in a new war between the United States and Russia. The president of the United States, Lyndon B. Johnson, kept in close touch with the events of the evening in case there was trouble. Governors in several states called out National Guard troops. And the Air Force checked and rechecked radar screens to see if there was an attack coming from Russia.

Soon the military had reassuring news: There was no attack. The statement allayed fears of war, but it didn't stop a flood of rumors. The event was too incredible for many to pass off as an accident. Machines broke down, but not this badly, so people argued that it had to have been caused by someone. But who? War protestors? Black revolutionaries? A few people decided that the American government had done it purposely to see how people would react in an emergency. Others told stories about a secret weapon that went out of control and caused the blackout. One young New Yorker came up with the most extreme suggestion of all. He announced that the power failure had been caused by flying saucers! Again people laughed. But even this story, as wild as it seemed, couldn't be dismissed, because no one knew enough to be sure that it wasn't true!

After asking how, people asked how long before the lights came on again. Once more there were no definite answers. Only the hydroelectric plants could restart quickly. Generators that ran on steam turbines needed time. When the generators tripped out, the turbines were stopped, as well. Automatic pumps and controls that fed fuels into burners ceased, and the fires died. And without fires there was no steam. Water cooled. Before the turbines could start again, power companies had to get the fuels burning and the water reheated. But they had to be careful; rapid heating and cooling could cause turbine parts to crack. Also, as the cascade spread across the Northeast, it caused switches to open circuits automatically. Before the power could come on again, these switches—called *circuit breakers*—had to be closed. Operators could close some of them from central control rooms, but others had to be hand operated. Reaching and then closing them would take hours. Finally, some generators needed a little jolt of electricity to restart. But, of course, there wasn't any.

So it would take time before the power came on again. Until then, people had to cope with the darkness. Thirty million people had to find a way to live without electricity. Their fears persisted. What would the darkness mean to a divided people?

An extraordinary night began to unfold, a night in which only one light burned steadily for millions of people. In the clear autumn sky, a full moon glowed brightly.

8

"One Sweet Mess"

THE BLACKOUT COULDN'T HAVE COME AT A WORSE time. It was rush hour, the time when workers throughout the Northeast had finished work and were heading home. In the large cities, at any moment between 5:00 and 6:00 P.M., a million or more people were on the move. On normal evenings, there were traffic jams, packed elevators, overcrowded commuter trains, and several hundred planes landing and taking off at area airports. It was the time of day when people depended most on electricity. They needed it to travel in safety. But on the evening of November 9, the electricity was gone.

The loss was felt most intensely at the airports. Officials of New York City's two airports spent hours nervously biting their fingernails. When the power failed, dozens of planes were approaching the city and more were hovering overhead. With the electricity gone, the airport towers lost runway lights to guide the planes, radios with which to talk to them, and radar with which to see them. Passengers were left hanging above the earth and potential disaster.

In all, about five hundred flights were diverted away from

the Northeast that night. Some passengers bound for New York landed instead in Cleveland, Ohio. Others touched down on the island of Bermuda about one thousand miles away. Most of the flights, however, ended up just outside the blackout region. The many diverted planes and passengers created confusion at airports in these places. Worst hit was Newark Airport in New Jersey. Almost two hundred flights bound for New York City ended up at that airport. The extra people and planes created chaos in the terminals and on the field.

But air travel everywhere was disrupted. Many flights scheduled into and out of the blackout area never got off the ground. Passengers spent the night sleeping on planes that had nowhere to go. People across the country, as well as in other countries, suffered inconvenience, as air schedules became as useless as northeastern electric switches. It was, according to one official of the Federal Aviation Agency, "one sweet mess."

A mess, to be sure. But not a tragedy. No planes crashed. That night every person aboard every plane survived. They were tired and stranded far from home, but they made it.

The mess was sweet, indeed. And not just for air travelers. "One sweet mess" characterized the blackout better than any comment made that night. Everywhere disaster loomed but seldom struck. On the contrary, people found that once they had accepted the darkness, they could actually enjoy it.

Still, the mess was real enough. Highways, railroad stations, subways, sidewalks, stores, restaurants, office buildings—all became scenes of confusion. Cars barely moved, and many didn't move at all. Since gas pumps ran on electricity, people low on fuel had to park their cars and find other ways to get home. Trains above and below ground, most of which ran on electricity, rolled to a halt, leaving more than a million people miles from home. Elevators stopped and hundreds were trapped far above the streets. In stores and other public places, the darkness forced crowds of shoppers to grope through unfamiliar surroundings.

Every place became a little unfamiliar, even homes. In the blackout people found the electric devices they'd come to depend on useless. The American dream home, the all-electric home, had become as functional as a log cabin. One Queens, New York, home, a model of electric living, was an extreme example of what the blackout meant to most houses. In the Queens house, the doorbell didn't ring. The intercom was silent. The clocks stopped. The hair dryer went cold. The can opener no longer worked, nor did the electric toothbrushes, the carving knife, the electric blanket, the heat, the hot water, or the electric-eye garage door. That night, in 40-degree weather, the family had to cook outdoors over a bed of charcoal, a cooking method invented closer to the Stone Age than the Electric Age. For light, they used flashlights and another ancient device—the candle.

In the early days of electricity, people kept lanterns on hand in case of failures. But by 1965, few people did, and many were now stuck in total darkness. Others rummaged through drawers and cabinets, trying to find battery-powered lamps and flashlights they hadn't used in years. Some never found them; others discovered that they had the lamps but no batteries to make them work.

Most people managed to find some kind of light eventually. Anything that burned or shone at all was used. A Con Edison worker stuck his head out of a manhole and asked a CBS reporter for a match to help him see. A Vermont barber rolled his car up to the window of his shop and finished a haircut in the glare of his headlights. A thirteen-year-old student studied by the glow of a fireplace.

But the most commonly used light was the candle. Hundreds of thousands of them were burned that night. In New York City, the Hilton Hotel alone used thirty thousand. The rich and powerful depended on them just as much as poor people did. U Thant, then the secretary general of the United Nations, led a procession with a candle in each of his hands. He and other UN officials trudged down thirty-eight floors from their offices to the street. Newspapers and corporations also spent the night bathed in candlelight. Editors and reporters of the *New York Times* put together an edition of the newspaper with the help of candles. And at one church, an employee of the stockbroker firm Merrill Lynch, Pierce, Fenner & Smith walked in, plunked down a sizable donation, and left with a fistful of devotional candles.

The night was filled with difficulties. Yet people managed, for the most part, to get through it. Not that they weren't afraid at times. People worried about everything from Russians to riots to the darkness itself. But they didn't moan about what was gone. Even as they struggled with the confusion, they laughed at their situation. With their laughter, their fears lessened. Most people began to see the evening as a kind of adventure, and they were able to turn their fears into wonder. "People have a way of solving their own problems," said a New York City police officer during the blackout. "Life goes on."

People showed imagination in their solutions to some of the problems of the night. In Vermont, for instance, people unpacked their warming freezers and set their frozen food outside, where the temperature had dipped below 32 degrees Fahrenheit. Some commuters, finding the highways snarled, the airports closed, and the trains stalled, remembered that there was another way to travel, by water. One commuter who lived in the Hudson Valley called his wife—the telephones worked because the phone company had emergency generators—and had her pick him up in their power boat at a New York City pier. Throughout the region, farmers displayed ingenuity in getting their milking machines working. Some hooked their tractors to the machines to keep them running.

But most people found that the best way of getting through the night was not to rely solely on themselves. It was to turn to others. Almost everyone was ready and eager to help. Perhaps fear pushed people together at first, but united in common fears, they became close to one another. So close did people feel that for many the evening turned into the opposite of the disaster that had seemed inevitable. Instead, the blackout turned the Northeast into the biggest neighborhood in the world.

People simply got together to help one another make it through the long, dark night. Race, wealth, power, politics—none of the things that divided Americans seemed to matter. A black maintenance worker in a New York apartment building summed up the feeling most people had that night. After leading a white tenant up ten flights of stairs, she waved away a tip, saying, "It's okay, honey. Tonight everyone helps everyone."

People performed all kinds of jobs and services that night. Asked or unasked, they did what was needed. When traffic backed up in towns and cities throughout the region, people jumped off the sidewalks and played traffic cop. In Providence, Rhode Island, a fourteen-year-old boy kept cars moving at one busy intersection. Businessmen helped untangle a terrible traffic jam in Toronto. In New York City, every busy corner was served by a citizen traffic cop. Actor Anthony Perkins played the role at one intersection, while a banker waved cars through at another. A Franciscan monk, dressed in a brown cassock, helped out at a third corner.

Northeasterners trusted one another and were seldom disappointed. In one instance, people from a stranded train followed a man with a flashlight. A woman in the group suddenly turned to her husband and asked, "Where are we going?" He answered, "We're just following the guy with the light. I hope he's not some kind of nut headed for a cliff." He wasn't; he led them to safety.

In a number of cities, groups of people organized spontaneously to help others. Some stood out on the streets and watched shops for signs of trouble. In Albany, New York, teenagers with portable radios went through apartment buildings, bringing news and instructions to tenants. Businesses helped people, too. Famous restaurants opened their doors and served soup, coffee, or drinks, free to anyone who happened by. Department stores let customers and workers stay the night. At one store in New York City, B. Altman and Co., five hundred people ate caviar and other exotic foods from the store's Imported Delicacies department. And the famed Manhattan movie theater, Radio City Music Hall, stayed open for stranded patrons. Three hundred of them decided to spend the night at the theater.

People were especially helpful and considerate to those who couldn't manage on their own. The young and the healthy could climb stairs, walk miles, or sleep on an escalator if they had to. But the elderly and the handicapped could not.

71

Many of them were caught outside in the dark and reacted with near-panic. They stood helplessly on street corners, some in tears. But usually strangers assisted them. In a number of instances, young people stopped cars and asked drivers to give rides to those who truly needed them. Few drivers refused.

As Tuesday night wore on, people began feeling good about one another, and they began to celebrate. Blackout parties sprouted up everywhere, from the Toronto airport to the Long Island suburbs, from the bars of New York City to the campus of Syracuse University. "We went out onto our street," one Long Islander recalled, "and there were other people out there. And soon everybody in the neighborhood was out there. Then someone brought out some food and someone else offered us something to drink. Before long we were just having fun."

Strangers became instant friends that night. They had the blackout in common and needed no other reason to start a friendship. In cities where people rarely know their neighbors, people spoke and shared with any stranger.

Goodwill and friendship thrived most among those who were stranded for the night. Thousands of commuters found themselves stuck on trains going nowhere. They talked to one another about the blackout, about the food they were missing, about themselves. They shared the few snacks they happened to have and some of them tried to play cards by the light of the moon. A few sang songs and others slept, their heads resting on the shoulders of people who had been strangers to them when their ride began. Most kept their sense of humor all night long. One man, for example, broke up fellow passengers when, after sitting three hours, he wondered aloud if he was on the right train.

As many as one hundred thousand commuters never made it on to a train that night. Because hotel rooms were taken within minutes after the power failed, they spent the night wherever they could find space. Many packed into churches. St. Patrick's Cathedral in New York City, one of the nation's largest churches, was filled with stranded people. Others settled in at the railroad stations. Over eighty thousand slept in New York's two stations, Grand Central and Pennsylvania. People curled up on the floors, on benches, on stalled escalators, and on stairs. And they, too, shared and spoke with all the strangers around them.

The subway riders of the Northeast were stranded in far more dangerous and fearful places. Over twelve thousand Toronto riders spent part of the evening trapped in darkened tunnels below the ground. And in New York City, stranded riders numbered eight hundred thousand. Some of them were marooned in tunnels below the city's rivers, miles from the nearest exit.

New York subways can be frightening places even when the lights are on. With the lights off, they are terrifying. But there, once again, people turned to one another, and that made the darkness a little less scary. In no instance was there panic. There was only cooperation. On one subway train, a woman fainted, but the word passed from person to person until someone came forward with smelling salts to revive her. Even in the pitch-dark tunnels, spirits stayed high. Passengers of one car had a party when a woman announced it was her birthday and unwrapped a cake. When a conductor of another train asked passengers how they were doing, they answered him in a single voice. "Fine!" they shouted.

Most of the eight hundred thousand riders got to the street within a couple of hours. In an orderly fashion, they marched out of the trains and along the tracks to stations or emergency exits. Usually conductors and police organized the processions to the streets. But sometimes people were led out by fellow passengers. In one instance, a blind woman, to whom the darkness was a way of life, guided the seeing to safety.

Some New York City subway passengers were trapped for several hours. Seventeen hundred found themselves stuck on a precarious railroad bridge over the city's East River. Police and subway employees helped them off slowly to avoid any mishaps. The rescue effort took seven hours, but it was wholly successful. Some riders chose to stay on trains until they started up again; they were afraid of going out onto the streets in a strange neighborhood. The New York transit au-

Woman passenger being helped from a subway tunnel during blackout Wide World Photos

thority didn't object and sent supplies of food and coffee down to them. Many of these passengers remained underground the whole night long because the subways didn't roll again until morning.

Unlike subway passengers, who usually had a way out, those stuck on elevators had none. They were trapped—totally—in small, dark cubicles that were often so crowded passengers could barely shift their feet without stepping on someone. Because their situation was worst of all, elevator riders grew closer than people stuck anywhere else. In one elevator, riders calmed a man who was worried about his heart condition, but in most, people just helped one another keep their humor and peace of mind. They played word games or talked about unlikely passengers for a stuck elevator. (One example: A Con Edison executive and any New York homemaker.) All were eventually rescued, as firefighters and maintenance workers cut through walls to get them out. In at least one elevator, the experience had lasting good effects. The passengers formed a Blackout Club to continue their friendship.

The goodwill so evident that night did more than make life easier for people; it helped save lives. People answered every call for help. For example, St. Vincent's Hospital in New York City made an urgent request for people to help pump iron lungs. The machines, which enable seriously ill people to breathe, had stopped and needed to be hand-pumped. When they heard of the problem, thirty patrons at a nearby restaurant rushed to the hospital.

Doctors, nurses, and other hospital employees also played heroic roles that night. Doctors delivered dozens of babies in the dark and performed delicate surgery by flashlight. Nurses walked through the wards quieting patients' fears and performing many important tasks. Other employees pitched in, too, and worked tirelessly with great success. Said Dr. Robert Gallance, who performed an intricate eye operation with just a battery-powered light, "You know, when things are ordinary around here, patients don't even get their bedpans. In a disaster, everything works out beautifully."

The spirit of the evening seemed to touch nearly every one of the thirty million people of the Northeast. Many had

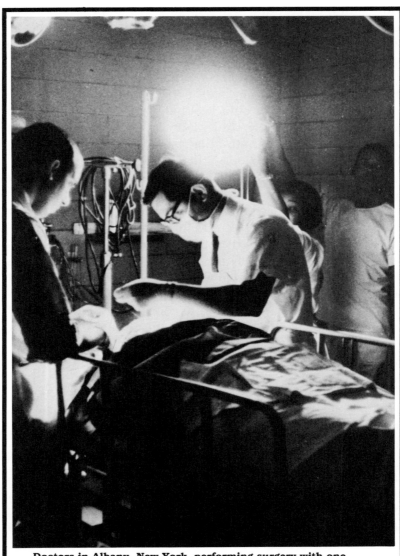

Doctors in Albany, New York, performing surgery with one emergency light United Press International Photo

feared that the blackout would lead to crime sprees and race riots, but there were none. In a couple of cities, a few exuberant (or drunk) individuals smashed some windows, and some trouble broke out inside the Walpole State Prison in Massachusetts. But there were no serious crime problems anywhere else. In city after city, police reported much *less* lawbreaking than usual. In New York City, authorities arrested only one-quarter as many people as they did on an average night. Though state governors had called out National Guard troops, the soldiers never had to battle anyone. Instead, most guardsmen spent the night directing traffic and taking emergency equipment to hospitals.

The night of Tuesday, November 9, 1965, was a night that could easily have been tragic but wasn't. And the reason is that people thought of themselves as part of a community in trouble, and not as individuals with their own problems. Those who lived through it would remember it always, and many would remember it as fondly as any experience in their lives. When it was finally over, one young woman said, "They should do this more often."

9

The End of the Night

SLOWLY THE DARKNESS BEGAN ROLLING BACK. Buffalo, New York, was one of the first cities to emerge from the blackout; power was restored there within an hour. Toronto was next, but it was out much longer. The whole city didn't have electricity until 8:30, more than three hours after the blackout started. During the next hour and a half, the darkness lifted through most of the Northeast. There was just one important exception. In New York City, the blackout lingered on. Some parts of the city didn't have electricity again until seven o'clock Wednesday morning, almost fourteen hours after the power failure had begun. But then it was over. Finally.

During the next day, millions shared memories of the night with friends. The question on everyone's lips Wednesday morning was: Where were you when the lights went out? Almost everyone had a story to tell.

Most of those stories were happy. But people learned that not everyone had had fun in the darkness. Hundreds of people had suffered broken limbs after tripping over household furniture or unseen obstructions on sidewalks. Cars had bumped into people on pitch-black streets, and though they had caused few major injuries because they were going so slowly, they sent many to the hospitals. In the worst instance, a bus had jumped the curb in the New York City borough of Queens. Nine people had been hospitalized in that incident alone. And several people had died because of the blackout. One, a visitor to New York City from Florida, had groped through the darkness in his hotel only to walk into an open elevator shaft. Others had suffered heart attacks after running up stairs or walking long distances to get home. Things could have been much worse, of course. But for the injured and their families, Black Tuesday was a dark and unhappy day.

Businesses, too, had suffered from the blackout. In supermarkets throughout the region, freezers had stopped freezing and foods had spoiled. In bakeries, ovens had cooled suddenly and thousands of loaves of bread had turned into masses of useless, half-cooked dough. Theater owners had faced as bad a night as they could have imagined. Movie projectors had slowed to a halt and shows had had to be canceled. Most live performances had been called off, as well. A few had gone on by the light of candles and flashlights, but not many people had shown up to watch. One play had been performed to an audience of seven people and two dogs. Of New York City's three morning newspapers, only the *New York Times* had been published that night, and it was just a few pages long.

Business continued to suffer on Wednesday. Many stores and offices, especially in New York City, closed early and some never even opened. Half the employees in New York didn't bother to go to work at all. The New York and American stock exchanges opened late and did very little business. Stockbrokers joked about investing in candle companies, but the slow day in America's financial center was no laughing matter. Business not carried on meant money not received, and that hurt companies dearly.

When all was said and done, the blackout turned out to be an expensive proposition. Companies lost millions of dollars. But taxpapers also paid a price. Most firefighters, police officers, and National Guardsmen worked the entire night and had to be paid overtime wages for their efforts. That cost brought the total price of the blackout to almost one hundred million dollars.

On Wednesday, few people gave much thought to the cost. Instead, they worried that there would be another blackout. And however much fun the first had been, the idea of a second was not very appealing. The possibility of another failure loomed largest over New York City. The first outage had damaged three of Con Edison's generators; protective devices on the machines simply didn't do their job. The utility admitted that supplies of electricity were "tight," but it tried to reassure the city. "We're still capable of producing all the power that's needed," company officials maintained.

But people weren't convinced. And with good reason. On Wednesday, no one, including power company employees, knew how the blackout had started. Since the companies didn't know how it had happened the first time, how could they be sure the same thing wouldn't knock out the power again? Northeasterners took nothing for granted. All day long on November 10, they prepared for another blackout. They bought every candle, flashlight, and lantern they could find. Some also bought portable radios, stoves, and other emergency equipment. Whatever the next evening brought, people intended to be ready.

There wasn't another blackout on Wednesday, though. Throughout the day and night, the power system held. It remained solid on Thursday. By then, people began to relax a little. They stashed their candles and lanterns and portable radios, and they returned to work and school. Life and the power system went back to normal. But deep down, a lot of people remained uneasy. Sure, the power system *seemed* to have returned to normal. But had it really? On Thursday, people still didn't know how the blackout had started.

Not that some people weren't trying to solve the mystery. Even before the blackout ended, President Johnson ordered the Federal Power Commission to find the solution. He told FPC chief Joseph C. Swidler to take charge of the investigation personally. After all, the government couldn't let something so disruptive just happen and remain a mystery.

Swidler sent investigators to the Northeast. They spoke

to power company officials and asked them the obvious question: Where did it start? But the officials could only say, "We didn't start it." Swidler even summoned power company heads to Washington to question them himself. But he got nowhere. At week's end, he confessed to reporters, "[We] may never trace just where it started." In other words, thirty million people could suffer a blackout for totally unknown reasons!

Swidler didn't give up, however, and through the weekend he pursued the investigation. At first, all he seemed able to do was disprove rumors. And the rumors were flying. One of them placed the start of the blackout at the Robert Moses plant. But even though it was hit soon after the trouble started, investigators decided the blackout hadn't begun there. Also, stories of sabotage cropped up again. Checks of equipment and power lines showed no evidence of this. Officials didn't really believe it, anyway, not because they had a better answer, but because at the time they didn't know how it could have been done. "If someone asked me to cause this thing to happen," said the chief engineer of one power company, "there isn't anything I can think of doing to cause it."

By disproving the rumors and by piecing together all the information he had, Swidler was finally able to figure out where it had begun. On Monday, he told reporters that the trouble had started "in the Ontario hydro-generating plants on the Niagara River." Exactly which one started it, however, or how it started, he couldn't say.

But at almost the same moment that Swidler made his comments, the Hydroelectric Power Commission of Ontario called a press conference. At the conference, the chairman of the Commission admitted that one of its plants had triggered the blackout. After a lengthy check of equipment, the Commission's engineers had located the exact cause. They found that a device the size of a telephone had failed. That little device had started it all.

The full story went like this: A little after 5:00 P.M., a power station in Syracuse made a call to the Ontario Hydro control center outside Toronto. The New Yorkers wanted more power. It wasn't an unusual request. With rush hour at hand,

Niagara River

The Sir Adam Beck hydroelectric plant, where the great blackout began Wide World Photos

83

cities often found themselves a little short. They frequently asked the hydroelectric stations to push more cheap power into the lines. On this particular evening, the plants of Ontario Hydro had power to spare. They told the New Yorkers not to worry. A controller reached for a dial to raise the voltage at the Sir Adam Beck Generating Plant Number 2, a Niagara River station. The technician did not turn the dial too far. Nothing should have gone wrong. But shortly after he touched the dial, the great blackout started.

Seven high-voltage lines were running out of the Beck plant. Five of the lines went west toward the cities of Toronto and Hamilton in Ontario. The other two crossed into the United States at the Niagara Gorge. At the time the controller raised the voltage, a total of 1.6 million kilowatts flowed through the lines of the plant, most of them heading west. The plant's sixteen generators were producing most of that. About five hundred thousand kilowatts were coming into the station from other parts of the grid. The load was typical; it certainly wasn't too high.

When the operator turned the dial, the voltage rose on all the lines, the ones going west as well as those headed east. And a *relay* on one of the western lines blew. The device was a type found on all high-voltage lines. If voltage goes above a certain level, equipment can be damaged. The job of the relay is to measure the amount of voltage in the line, and to shut down the line entirely if it goes too high.

That was exactly what the relay did to one of the lines coming out of the Beck station. It shouldn't have happened; the voltage *wasn't* too high. The relay just failed. When it did, the power flowing through the line shifted automatically to the other four lines running west. But this extra power was too much for those lines. Relays on them properly shut them down. In a matter of moments, no more power flowed west from the Beck station.

When electric current can't flow in one direction, it turns and goes in any direction it can. In this case, that direction was east. All the Beck plant's power—1.6 million kilowatts—suddenly crossed into the United States. The electricity crashed into the American power stations near Niagara Falls.

84

That was why the power level rose tremendously at first. The surge lasted, altogether, only five-sixths of a second, as equipment in both the United States and Canada quickly shut down. The loss to the pool was too great, however, and the blackout of '65 followed within minutes.

So one small relay did it. A system larger than most people can imagine was knocked out by one tiny device. The idea was both laughable and frightening. If one relay could knock out a whole power pool, how many other devices could do it, people asked. America's electric power supply suddenly appeared very vulnerable. Americans' faith in the electric switch, and in technology itself, was severely shaken. Wrote A. M. Rosenthal in the book *The Night the Lights Went Out*:

> We had all been told time and again that it could never happen. . . . We had all accepted that and yet it did happen. . . . Then all of a sudden they weren't telling us that it could not happen again. Suddenly we were being told by our scientific and engineering betters that it could indeed happen again, maybe right tonight. So now we believe that it could happen again and we have very little faith in our engineers, a good reaction indeed for all of us.

But if the blackout changed the views of some people, it didn't change the strategy of the electric companies. Though a big system had led to a big blackout, they called for an even bigger system. The government agreed. A Texas congressman, Walter Rogers, for example, talked about a "three-way" system. It would have required a back-up system to back up the back-up system that backed up the main system. Others talked again of a national grid as the answer to blackouts. Some people also advocated more advanced technology. They especially wanted to see the decision to cut away from the grid made by computers instead of human operators.

Studies and hearings into the '65 blackout went on for two years. Changes were made. Power company equipment and procedures were redesigned. Protective devices and grid interconnections were improved. Pooling arrangements were strengthened. In the Northeast, the informal pool, CANUSE, was reorganized and renamed the Northeast Power Coordinating Council. It was made up of four small, well-organized pools: Ontario Hydro, the New England Power Exchange (NE-PEX), the Maritime Power Pool in eastern Canada, and the New York Power Pool (NYPP). The NYPP, in fact, came into existence only months after the blackout of '65. Through the new pooling arrangements, companies hoped that the Northeast electric system would be better coordinated and managed. Yet people wondered if the changes had really made that much difference. Many people feared that there would be trouble with the Northeast electric system again.

The blackout of '65 passed into history. The night became an American legend. A movie appeared called *Where Were You When the Lights Went Out?* It starred Doris Day. Though hardly destined to become a film classic, it revived lots of memories. As with most legends, the memories were perhaps a little more satisfying than reality had been. But there can be no question that November 9, 1965, was a remarkable night, a unique night, one that probably never could be repeated again.

Unfortunately, the same could not be said of blackouts themselves. The darkness would return.

Aftermath of the 1977 blackout violence
Olivier Rebbot/Sygma

Part III

THE NIGHT
OF THE
"STICK-UP MAN"

10

The Exhaustible Supply

IN THE MONTHS IMMEDIATELY AFTER THE BLACKOUT of 1965, America seemed to suffer an epidemic of power failures. They had struck before, of course, but now they appeared to hit more often. And they were especially bad. The first big one came only a few weeks after November 9, on December 2. A fuel line that fed natural gas into a generating plant conked out, and the entire El Paso Electric Company system collapsed. The city of El Paso, Texas, nearby Juarez in Mexico, and many of the surrounding communities lost electricity for up to two hours. The president of El Paso Electric found the failure embarrassing. Only days before, a reporter had asked him to comment on the Northeast blackout, and he had replied confidently, "It can't happen here." On December 2 it did, and he sputtered in dismay, "It's unbelievable!"

It was hardly that. Four days later another city lost its electric power. This time it was Beaumont, Texas. The power went out for up to three hours when control equipment broke down.

And the epidemic grew. Dozens of blackouts struck in the next few months. If the November 9 blackout made Amer-

**Philadelphia man helping unsnarl traffic during blackout of
June 5, 1967** Wide World Photos

ica's electric system seem vulnerable, those that followed made it seem dangerously frail. Everything from wind to vandals knocked out whole sections of the country.

In 1966 someone tossed a wire across power lines, and 125,000 Alaskans lost their electricity for two hours. That same year 600,000 people in the Nebraska-Dakota region lost power for two and a half hours after a mechanical breakdown. A few months later a tree fell, and power to 295,000 people in Missouri and Arkansas blinked out for an hour. A large section of Virginia went out for a time, and so did chunks of California, Ohio, the Pacific Northwest, and several states along the Gulf of Mexico. From 1965 to 1967, nearly every state and every region experienced one or more large-scale blackouts. Americans wondered if the Electric Age was giving way to a new age, the Age of Blackouts.

On June 5, 1967, it seemed certain. On that day, for the third time in five years, a whole region went out. The victims this time were the thirteen million people served by the P-J-M pool. Once again, power companies let a cascade develop. Operators caused this one. They made a mistake and overloaded a 230,000-volt transmission line. As a result, the line shut down and the cascade began, shutting off power from Newark, New Jersey, to Wilmington, Delaware. Some parts of the system were out for as much as ten hours.

The failure started at 10:15 on a bright morning. At that hour, most people were indoors at work or school. And the sunlight and warm weather made the loss seem minor compared with the loss in '65. There were many of the then-familiar stories of traffic tie-ups, stuck elevators, and acts of heroism, but in at least one other respect, this blackout was different from the one two years earlier. The spirit of '65 was missing. There were no celebrations; people mostly felt inconvenienced and annoyed. Philadelphia subway riders, for example, came out of the ground with none of the humorous little stories their New York counterparts had to tell. When asked how things were in the tunnels, one rider answered bluntly, "Very dark."

Once again, a power pool had failed. People throughout the country were worried. *Newsweek* magazine asked the question on the minds of many Americans: "How many more times [will] sections of the country have to grope in the dark?" To some people, the answer already was clear: many, many more.

The power companies weren't giving up on trying to improve their systems. If anything, the bad publicity they got from every failure made them redouble their efforts. They were rapidly losing the trust of the American people, and they sought ways of restoring it. Again they spoke of adding new machines, of creating the still-unrealized national grid, and of other ways to make systems better by making them bigger.

Whether they got better or not, power systems had to get bigger. Americans, though worried about the reliability of electric power, continued to use more and more of it. From 1960 to 1970, consumption of electricity doubled in the United States. People had come to believe that the supply of electric power was practically without limit, and nothing that had happened changed that belief. True, the power companies weren't perfect. But they provided all the power people wanted, and they provided it cheaply.

Most Americans wanted to see electric power production grow. Most Americans; not all. In the 1960s, groups were beginning to question the rapid, unchecked growth of electric systems. They claimed that the country paid a price for that power—a price not measured in dollars and cents. The groups were not actually opposed to the growth of power; they just wanted it to be more carefully planned and controlled.

The utilities, opponents claimed, were damaging the basics of life. They were hurting the air people breathed, the water they drank, and the land on which they lived. The critics said our environment was being destroyed, polluted, and not just by the electric companies. Other industries and even automobiles did some of the damage. But the dark clouds of smoke billowing out of the chimneys of the generating stations couldn't be ignored. Every generating plant that burned fuels like coal and oil was a polluter. And pollution, the protestors claimed, was a danger to every person in the country.

Just how serious a danger it was, was not widely appreciated before the 1960s. But as pollution grew worse, Americans began hearing chilling stories. They learned that the soot and smoke in the air could lead to lung diseases and even death, and that chemicals in the smoke came down in the rain and into the country's drinking water.

When people saw that pollution could affect their health, they grew concerned. The environmentalists took the leading role in the fight against pollution. But soon many people joined them. Groups held demonstrations and demanded that new laws be passed. They wanted the government to prohibit cars that spewed out polluting exhaust. They called on officials to make sure that factories didn't send dirty smoke into the air or dump chemicals into rivers and other waterways. And they demanded that the government take steps to stop pollution by the electric industry.

Utilities were criticized not just for polluting the air, but also for building nuclear-powered generating plants. Nuclear-fueled plants had at one time been the great hope of the electric industry. Companies had proposed building hundreds of them across the nation to provide more cheap power to American consumers. But after a few were built, environmentalists turned against them.

Nuclear plants produce no air pollution; nothing in them is burned. So why did the environmentalists want to block them? Because nuclear plants can endanger life far more than any other kind of plant. We noted in Chapter 1 that protons are bound in the centers of atoms and only great forces can free them. But in some atoms, the centers, or *nuclei*, are unstable. On their own, they release protons and other particles. The release, however, is accompanied by the release of energy, great quantities of energy. The energy is, in part, in the form of heat. That heat can boil water to produce steam to run turbines. But as the atoms release energy, they change into other unstable atoms, some of which emit particles constantly for years. Those particles, called *radioactivity*, are extremely dangerous. Scientists know that human beings exposed to fairly small amounts of it will develop cancer. A person who is exposed to high doses of radioactivity may die within a matter of minutes. If large amounts of radioactive particles escape into the air, they may fall over a wide area, bringing disease and death with them.

For the most part, all radioactivity in nuclear power plants stays within the unit that houses the nuclear fuel, called the *reactor*. But water and steam come in contact with the fuel, and they become radioactive. And all the equipment in a reactor building becomes radioactive, as well. Of course, every reactor has safeguards to prevent radioactivity from leaking out. But what if something went wrong and it did leak out? What if even small amounts of steam seeped out through a hole in a pipe? That's what environmentalists feared, and that was why they fought to stop the building of every nuclear power plant the utilities proposed.

City, state, and federal governments involved themselves in the environmental issues of the 1960s. To a great extent, they supported the environmentalists. Officials didn't go as far as to prohibit the building of nuclear plants, nor did they close down plants that were polluting the air. But they did pass regulations that made the building of new power plants more difficult. Soon they reserved the right to approve the location of power stations, the fuel to be used, and the equipment to be installed. Governments also forced utilities to show that new plants wouldn't damage the environment. Furthermore, they required companies to install clean-up equipment in old plants so that they would produce as little pollution as possible.

The new rules required power companies to make studies, draw up plans, and hold hearings before they could start building new facilities. Often government agencies rejected plans several times before construction could begin. Sometimes companies had to spend years studying and planning a project. Then more years passed while the plant was being built and outfitted. Still more years went by before the plant was able to begin operation. The whole process, especially if the plant was nuclear, took ten or more years in some cases. And it often cost utilities millions of dollars beyond the cost of construction.

The environmental movement focused attention on some important problems. But because of it, a different kind of problem began to develop. While companies had trouble building new plants, Americans kept using and wanting more power. It was only a matter of time before there would be shortages.

Sure enough, they started before the end of the 1960s. New York City, which had adopted strict laws against pollution, was again the center of a power crisis. By 1969, the supplies of electricity in the city were low. During hot summer days of that year, when air conditioners and refrigerators drew extra power, Con Edison's facilities were pushed to their very limits. Even with help from the grid, the city barely pulled through without a major blackout. In fact, Con Edison frequently didn't have enough electricity to meet demand. But it got through the summer with the help of a process called *load-shedding.*

The idea behind load-shedding is simple: If the demand on a system becomes too great, part of the system is cut out to save the rest of it. The first step in the process is for a company to reduce voltage slightly throughout its entire system. As a result, every appliance, every light, every machine gets less power. And with less power being used, demand on the generators decreases. A small reduction in voltage—5 percent or less—isn't noticeable, except in machines with very sensitive motors. If a 5-percent reduction in voltage isn't enough, a utility will reduce it further, up to 8 percent. An 8-percent reduction is noticeable. Television pictures shrink and other electrical appliances begin to sound strange and work improperly. The reduction of voltage is often called a *brownout.*

If a utility still can't meet demand after an 8-percent reduction of voltage, it will begin to black out small areas to reduce load. A company, faced with an imminent blackout, will knock out power to a street and then perhaps to a neighborhood or two. If demand is still too great, a utility might black out a couple of towns to save the rest of its system.

In the 1960s, Con Edison designed its system so that it could, if necessary, shed up to half its load to save the other half. Controllers had switches with which to do it by hand. But Con Ed's load-shedding usually was done automatically by computer. Computers operate almost as fast as electricity travels and can spot problems and act on them before a person can. So computers began to play a greater role in electric power in the 1960s.

During the summer of 1969, Con Edison's load-shedding equipment was tested repeatedly. For days, the city hovered on the edge of a blackout. The power shortages were so bad that Con Edison officials begged people to cut down on their use of electricity. But while the city came to the brink of darkness again and again, it never slipped into it.

Con Edison struggled through shortages for the next few years, and it didn't have a single blackout. Then, largely because of the addition of a 900,000-kilowatt nuclear-powered generator to its station at Indian Point, New York, Con Edison stopped struggling and managed to get through summers without power crises. But the basic problem hadn't ended. With the new environmental rules still making things tough for the electric industry, other shortages cropped up. Old equipment broke down, and there was nothing new to take its place. In a part of the south, equipment problems at one point fostered severe shortages that led to weeks of brownouts and blackouts.

In the early 1970s, the future of electric power suddenly seemed uncertain. Companies worried that for the first time in decades they wouldn't be able to follow the old plan—provide more power to more people for less money. Although they were concerned, the companies didn't change their approach. They still thought that bigger would result in better. And Americans used more power than ever.

The environmental rules made it hard for utilities to pursue the old idea, but something else that began in the 1970s made it even harder. It went by the now-familiar words *energy crisis*. The energy crisis developed because Americans used tremendous quantities of fuels, particularly oil and natural gas, to run power plants, cars, airplanes, factories, and so on. People used so much oil and gas that America couldn't produce enough by itself. We had to buy fuels from other countries. For a while, there were no problems; fuels were plentiful and cheap. But then the countries that sold the fuel realized how much this country needed what they had. They understood that they could, first, charge whatever they wanted, and, second, stop selling to America if they didn't like this country's policies. They did both. They raised the price of oil and gas and, in 1973—for a while, at least—several countries stopped selling oil to the United States.

The energy crisis affected the electric power industry from its start. It scared the utilities greatly. Many of the nation's power plants ran on oil; utilities used more than one million barrels—forty-two million gallons—of oil per day. If the companies didn't get fuel there would be power shortages, and power companies would be able to do nothing. Whether America got enough fuel was up to other countries. At the whim of others, our nation could face shortages and blackouts.

Though threatened, there were no cutoffs of oil after 1973. Still, the energy crisis had a real and ongoing effect on electric power. There was oil, but it was expensive. The plan—more power to more people for less money—seemed like a joke in the 1970s. Rising prices for fuel destroyed the idea of cheap power, and electric charges soared. In a few years, prices doubled, tripled, and quadrupled. No one was happy about it, but no one was doing much about it, either.

Both consumers and companies dealt with the new situation mostly by blaming others. Companies blamed the government for establishing rules that kept them from building nuclear plants, which, they said, would create more cheap power. And the utilities blamed the oil-producing countries for raising the price of fuel, which made power so expensive.

Consumers also blamed the government and the oil producers. But, most of all, they blamed the utilities. They wondered if the companies weren't just using the energy crisis and the new environmental rules to make more money. Consumers felt betrayed, cheated. Sold on the idea of an endless supply of cheap power, Americans had grown accustomed to using it. People could not even imagine using less. Although the rate of growth did slow down, our use of electricity continued to grow. It was much easier to blame the utilities than to change habits. Perhaps utilities should have foreseen the skyrocketing prices. The power companies protested that they could not have predicted the increase. But Americans resented them for it, anyway.

With people feeling distrustful, the last thing utilities wanted were blackouts. But they struck. They weren't as bad as the ones in '62, '65, or '67, but the entire state of Utah did go dark for a time in 1976, and a year later, two million people in Florida spent as much as four hours without electricity.

People were no longer surprised to learn about widespread blackouts. The hated power companies always messed up, they thought. And the problems the blackouts created were the usual ones. All power failures seemed to be the same.

But on July 13, 1977, a new twist was added to the story of blackouts. It was a terrible twist indeed.

11

Wednesday, July 13, 1977

THE YEAR 1977 STARTED AS BADLY AS ANY IN memory. Droughts ravaged the western states, particularly California. Water was so scarce that some California towns limited the amount that people could use. Swimming pools stayed dry and lawns turned brown. Fires swept through dried brush and forest lands throughout the state. The ground was so parched it seemed that half the West Coast could burn.

In the Midwest and East, meanwhile, the winter was the coldest on record. During the month of January, the temperature seldom rose above 32 degrees Fahrenheit in about half the states in America. In some, the frigid weather was much more severe. In the Midwest, especially, weeks passed with the thermometer hovering at or below zero. People longed for the relief spring and summer would bring.

But the summer turned out to be the flipside of winter. After months of cold, the same states sweltered in unbearable heat. Temperatures soared above 100 degrees on many days, and the heat seldom broke. Each day seemed as bad as the one before it.

Weather forecasters were puzzled. What did the extremes of temperature mean? For the morning of July 14, the *New York Times* planned a feature article about the weather. However, for reasons we'll soon see, only part of the article appeared that day.

Few places suffered a more uncomfortable summer than New York City. During the day, the hot sun baked the city's streets. The concrete and tar heated up like coals in a fire, so that anyone walking along felt the heat from both above and below. Added to that was high humidity, muggy, heavy air that made even breathing uncomfortable. Being outside in New York City that summer was like walking through a steam bath. Outdoors, only the beaches were bearable. And inside, temperatures were no more tolerable without air conditioning.

The days were bad, and the nights were little better. If anything, people found the nighttime heat less acceptable. They were willing to put up with hot days, as long as they had some relief at night. But there wasn't any relief. A thermometer showing 100 degrees during the day would fall at night only to the nineties or eighties. Worse, apartments and houses seemed to hang on to the daytime heat all night long. Air conditioners offered the only comfort. Those who didn't have them suffered continuously.

And none suffered more than the poor who couldn't afford air conditioners. They had no escape. The only place for them to go at night was out onto the streets. True, it was hot out there, and it was also dangerous, as poor neighborhoods had the most street crime. Yet the streets were more bearable than the apartments. Night after night, the poor went outside until it was time to go in and try to sleep.

New York City's poor suffered from more than the heat that summer. A record number of the black and Hispanic poor were out of work. Three out of every four young blacks and Hispanics had nothing to do and nowhere to go during the day or night. Most of them just "hung out," and, while they did, they grew angry. As the heat wave went on and they and their families suffered through it, they grew even angrier. The heat and the lack of jobs were a dangerous combination;

together they became a kind of bomb. If the right event occurred, that bomb would surely explode.

July 13 was like most days New Yorkers had endured that summer. The thermometer climbed to the mid-nineties during the day. By evening, it still registered in the upper eighties. It was also extremely humid. No one wanted to be outside, but the poor had nowhere else to go. As the sun set, they poured out onto the streets of their neighborhoods and tried to cool off.

The heat wave was putting a strain on the city's power supplies. Air conditioners throughout the city ran nonstop. Every freezer and refrigerator gobbled power to keep food cold, forcing Con Edison to push huge amounts of power into the lines. The city and Westchester County were drawing almost six million kilowatts of power that evening. Still, the utility wasn't having any real problem meeting that demand. It had seen worse days; in the past, peak load had reached eight million kilowatts. Con Edison didn't even need to shed load. It had power to spare.

Con Edison had eleven generators in operation that night. However, it was turning out only half of the six million kilowatts. The utility was importing through the grid a massive total of almost three million kilowatts. (Remember that in 1965 they were importing only 350,000.) Con Edison's links with the outside were important. Without them, the city's power system would have great difficulty meeting demand.

Electricity flowed into the Con Ed system from three directions: from upstate New York to the north, from the P-J-M pool to the west, and from the Long Island power company to the east. Most important were the lines that came down from the north. They carried a huge river of electricity into the ocean of need that was New York City. They were particularly important to the city that summer because one of the lines tying Con Ed to the P-J-M pool was out of service. However, Con Edison had little reason to fear that anything would go wrong the night of July 13. All the remaining transmission lines were working perfectly.

Then thunderstorms appeared over Westchester. The

storms were unusually bad. Lightning ripped the skies and thunder boomed over the suburban county. Since power lines have static wires to protect them from lightning, Con Edison didn't worry at first. But at 8:37 P.M., a single bolt of lightning pierced the dark, avoided the static wires, and struck a 345,000-volt transmission line near Buchanan, New York. The line carried power from the Con Edison nuclear plant at Indian Point, thirty-five miles north of the city. When the lightning struck, current surged through the line and protective devices automatically shut it down. Power flowing out of Indian Point now had nowhere to go. The generators inside the plant sensed that and quietly stopped themselves. Nothing had actually been damaged; lightning had just fooled the equipment. But the joke was on Con Edison, as the output from Indian Point went off the grid. Nine hundred thousand kilowatts of power disappeared from the system.

The loss of Indian Point was hardly good news, but it wasn't a disaster, either. Not yet, anyway. Con Ed managed to fill in for the lost power by pushing other generators harder and by getting more power from the grid. In a short time, Con Edison's controllers had the situation in hand.

But the thunderstorms still raged over Westchester. At 8:55 P.M., lightning struck again. This time one of the transmission lines that came down from further upstate went out. The loss was serious, and the possibility of a blackout loomed. Con Ed's computers began to take action. They began to shed load.

First, they cut voltage to the entire system by 8 percent, but that wasn't enough. They began blacking out small areas of Westchester. If the computers had performed properly, the rest of the system might have been saved and those blackouts wouldn't have been serious. Power in those areas would have come back on in a few minutes.

Then, at 9:19, for unknown reasons, more lines from upstate shut themselves down. The situation was now truly desperate for Con Edison. Unless half the system was shed, New York City and Westchester faced a complete blackout. To their astonishment, Con Edison's operators found that the computers weren't performing properly and couldn't shed half the

**Severe lightning storms near New York City on the night of
July 13, 1977** Wide World Photos

load. The automatic controls themselves could cut out only about a third of the system. Quickly, the operators began turning every dial and throwing every switch in sight, trying to cut the rest manually. It was a race against time, and with every passing minute, they knew they were losing.

At 9:29, the race ended. The city had been drawing more and more power through its remaining ties to the grid, but it had been drawing too much. The utilities near New York City were getting themselves into trouble. The Long Island Lighting Company (LILCO) came very close to overloading its system as it tried to help Con Ed. To save itself, it broke the transmission line that linked it to Con Edison. No more power reached the city from the east. At almost the same moment, the connection to the west opened, too. The P-J-M pool wasn't able to fill in for Con Ed's losses alone. That left the New York City utility on its own. At this point, it probably couldn't have kept going even if its load-shedding equipment had been working. Without it, Con Edison's system was doomed.

As each tie to the grid went out, Con Ed's generators pushed harder and harder. They couldn't fill in. Soon they neared the point of overload, and their own protective devices took over. First, the huge generator known as Big Allis tripped out. Then, all over the city, other generators followed. By 9:40, New York City and Westchester County had faded into the night.

Some utility people congratulated themselves. They had prevented Con Edison's troubles from leading the rest of the Northeast into the darkness. The grid had worked, they said. It was now able to protect itself, to single out trouble and cut it off. Of course, that hadn't been the original idea behind the grid. It wasn't merely supposed to save itself. It was supposed to prevent *anyone* from suffering a blackout. On the night of July 13, 1977, nine million people were left in the dark. It was not the most extensive blackout to date, but in only a matter of minutes, people would realize it was by far the worst.

12

The Nightmare

WHEN THE LIGHTS WENT OUT, NEW YORK CITY'S mayor, Abraham Beame, was speaking to an audience in a synagogue in the Bronx. He thought a fuse had blown, but he suggested that Con Edison had shut off the electricity on purpose. "See what you get for not paying your bill," he joked. It was probably the last time the mayor laughed that night. The night was a replay of 1965 only in that the lights went out. The other blackout had been a celebration; this one was a disaster.

During the night of July 13, 1977, thousands of New York City's residents changed. They forgot that people live by certain rules, by certain standards of behavior toward one another. It was as though the rules had somehow vanished with the lights. Under the cover of darkness, people went on a furious rampage of looting and destruction. As the night wore on, the violence spread, and the city echoed with the sounds of police sirens and breaking glass. It was a night of terror unlike any the city had seen before.

Of course, not everyone in New York rampaged. Nor were most people terrorized by those who did. The majority of New

Yorkers spent a peaceful night. The closest most people came to the violence was to hear about it on their radios. Many people tried to re-create the good feelings of 1965, but the events going on all over the city made enjoyment of the darkness next to impossible. "It really brought you down to, you know, hear about all of that violence," said one young man. And comparing '65 to '77, another person shook her head and sighed, "What a difference!"

Yes, it was different from '65 and not just because of the violence. The weather was also different, hot, while in 1965 it had been cool. In November, 1965, most people wanted to get home quickly and find a warm place to sit down. Twelve years later, in the sticky heat, homes were even less comfortable than the outdoors. With air conditioners stopped by the power failure, homes grew hot and stuffy before long. As for the poor, they were already out on the streets in great numbers when the blackout struck.

The time of day was different, too. This blackout started at 9:30, after owners had closed their stores and left for the night. In 1965, the power failure hit a little after 5:00 P.M. Most storekeepers were still in their shops. With the power gone, they merely closed up and stood guard over their property. This time, most stores were guarded only by metal gates that could be pried open, locks that could be smashed apart, and alarms that had been silenced by the power failure.

Most important, the way people felt about the blackout was different. When the streets went dark in 1965, no one knew how or why it had happened, or how long it would last. Most people were afraid of nuclear war or some kind of violent uprising. That common fear brought people together. In 1977, however, everyone assumed that the blackout was due to a mechanical failure. No one feared war or mad bombers. Nearly all New Yorkers believed that Con Edison had simply let the electric system fall apart again.

Finally, in 1965, people felt the lights might come on again at any moment, certainly within a couple of hours. On the night of July 13, 1977, people were sure the lights would stay off a long time. Most figured they'd be off the whole night.

Anyone thinking of taking advantage of the darkness knew he had hours to operate.

All five boroughs of New York City suffered some violence during the night. But in a couple of them, the amount was slight. Staten Island, largely a middle-class residential area, experienced the least damage. A few windows were broken and some cigarettes were stolen. Queens, the city's largest borough, wasn't hurt badly, either. In all, less than one hundred Queens businesses sustained damage that night.

But in the other three boroughs, the story was different. Each experienced widespread looting and vandalism. In Manhattan, about three hundred shops and stores were robbed and damaged. Most were in the black and Hispanic communities, particularly Harlem, neighboring Spanish Harlem, and the upper West Side. Vandals struck in more than half a dozen other neighborhoods from one end of Manhattan to the other.

Still, the violence in Manhattan was slight compared with that in parts of Brooklyn and the Bronx. In both boroughs, people of the poor neighborhoods seemed possessed by a kind of madness. Almost everyone who went out got caught up in it. Criminals and noncriminals joined in a frenzy of lawlessness. They wrecked whole streets. On some, they looted every single store, and when they couldn't find anything more to take, they tore the stores to pieces. At times, they even lit them on fire.

For reasons we perhaps can never fully understand, accepted ideas about right and wrong disappeared. People operated by their own rules, rules suiting the moment. "We're doing right," one young ghetto resident insisted at the height of the violence. "There's no real big thing about it."

The trouble started a few minutes after the lights went out. The people who began it were mainly hardened criminals, men and women with long police records. Most of them didn't need any excuse to break into a store; they did it regularly. When they saw the lights go out, they understood they had a special opportunity. Quickly, they moved to take advantage of the situation. Some just picked up garbage cans or any other heavy objects and threw them through the nearest store windows. Not having prepared themselves for this opportunity, the first group of robbers merely stuffed their pockets and went home.

Others did a more thorough job. They got cars or trucks, broke into stores, and then filled their vehicles with loot. One man with a long criminal record saw what was going on and persuaded a few of his friends to help him. Together, they wheeled a thirty-eight-foot van out onto the streets. Before the night ended, they loaded it with everything from color TVs to leather jackets.

For the first hour or two after the blackout began, the professional criminals ruled the streets. They came out so swiftly and in such great numbers that the police couldn't stop them. There simply weren't enough officers on duty. Calls, of course, went out to all off-duty police, but they were slow in reporting to their stations. Because of the blackout, the officers couldn't get to a station except by automobile. And traffic, as usual in a blackout, was a mess. Until the officers reported for duty and went out onto the streets, the criminals had little opposition.

The professional crooks cared only about getting as much as they could for themselves, but they did bring other people into the action. They convinced the angry young ghettoites that looting and destroying was a "cool" thing to do. The criminals even organized the young people. They got them to help pry open protective gates and break locks. Naturally, the professionals made sure they took the best merchandise for themselves. At times, they even kept young looters out of shops at gunpoint until they had the quality items tucked away.

When the criminals finished, they invited their helpers to take what was left. There was seldom much of value remaining, but, infected by the wild spirit of the evening, the young people went through the stores eagerly. After they stole, they broke things. To many young people, the evening became a "fun" adventure. One young woman even called the rampage "sort of beautiful." But another ghetto resident was probably closer to the truth when he noted, "It's a chance to let our frustrations out."

Seeing that the police hadn't made many arrests of either the criminals or the young people, more and more people joined in the "fun." Even working people jumped in. Soon the streets of some neighborhoods were packed with people. In the Crown Heights and Bushwick sections of Brooklyn, two of the worst-hit neighborhoods, it was almost impossible to get from one side of the street to the other. Thousands crowded into a few hundred feet of space.

Not everyone in those neighborhoods took part in the looting. Some just stood and watched. But a huge number of people did get caught up in the action. They acted like scavengers as they picked through the stores already left a shambles by the criminals and the young. One woman, for example, was spotted down on the street, sifting through the broken glass outside a jewelry store, hoping the first thieves had dropped a gem as they made their escape.

People weren't interested only in stealing jewels that night. They took just about everything, from Pontiacs to Pampers, from appliances to sneakers. Particularly sneakers, which were among the most sought-after prizes of the night. Shoe stores and sporting goods stores selling sneakers were hit early and completely. The next day, finding a pair of sneakers in a store in any poor neighborhood was nearly impossible. Looters, however, did a brisk business in stolen footwear. For days after the blackout, Pro Keds went for about five dollars a pair in back rooms and on street corners.

Looters not only took everything; they also took from everyone. They stole from shops owned by men and from shops owned by women. They stole from stores owned by old people and those owned by young people. They stole from bystanders, and they stole from one another. They stole from white people, and they stole from blacks and Hispanics, too. During the race riots of the 1960s, mobs had protected black-owned stores even as they were destroying those owned by whites. On the night of July 13, however, the mobs cared little about race. The black owner of a Harlem sporting goods store

111

appeared on the scene while people were robbing his place; it didn't matter. He even recognized some of the looters. But they didn't stop, and there wasn't much he could do to stop them. He could only stand and watch as they turned his store into a ruin.

Actually, the black- and Hispanic-owned stores were often hit *first* because they were the easiest targets. Some large, white-owned chain stores were locked up so tightly nothing could get them open. Others had armed guards on the scene. Small, black-owned businesses, on the other hand, had weak doors and gates and no one on the premises. So on the upper West Side, while a large chain department store suffered no damage at all, a small clothing store and a jewelry store were wiped out totally. Both had minority owners, but that meant nothing. Said one saddened black businessman in Harlem, "We thought we were part of the community. We were wrong."

In the worst-hit neighborhoods, the only small stores to escape damage were those guarded by their owners. Ernie Blye, the black proprietor of a Brooklyn tailor shop, stood watch over his store all night. He kept the mob away because he had a gun, a dog, and ten cans of potash, which can blind a person. A black druggist used a gun to hold off looters for a time, but they finally got to his store. He made the mistake of leaving for just a few minutes, and he came back to find that he had been gone too long. His store was a wreck.

By the early morning, most of the people left on the streets were the scavengers, the last group to get involved with the looting. It was their bad luck that at just that time, the police were ready to regain control of the streets.

It wasn't an easy task. Never in the city's history had the authorities faced so much trouble in so many different places at once. But from around midnight on, the police were ready to move in on the looters in force. They swarmed in on the mobs and arrested people by the dozens. Prisons throughout the city soon overflowed. One prison, called the Tombs, had been closed in 1974 because of poor conditions. Because of the huge numbers arrested during the night, the city decided to reopen the infamous jail. The other city prisons couldn't hold any more people.

Before sunrise on Thursday morning, July 14, the police had taken control of most of the city neighborhoods. But in a few, notably Crown Heights and Bushwick, the mobs still ruled. The police tried repeatedly to drive them away or arrest them, but the authorities always seemed to be just a step behind the mob. They would move in on a group of looters at one store, grab a few, and arrest them. In the meantime, the looters who escaped would go hit another store someplace else. The police couldn't catch enough of them to make a difference.

The blackout itself was the main reason the police couldn't act fast enough. The failure had knocked out the police car radio system. Squad cars couldn't talk to stations, so they couldn't get organized to act together. Police on foot had walkie-talkies, but these have a short range. Messages sent over them couldn't reach enough officers in time for them to catch the groups of looters.

Also, police cars began running out of gas, and without electricity, pumps didn't work, so they couldn't get refills. Most police officers were forced to keep up with looters on foot. In the worst neighborhoods, they couldn't manage it.

The police showed great patience that night in dealing with looters. Though they had trouble getting the looters off the streets, the police never took rash action. If they failed to catch a group, they tried again. They never let themselves get so frustrated that they used excessive violence. Though they might have been tempted to fire their guns, they kept them holstered.

Despite all the lawlessness, few people were killed that night. Two died as a result of the looting. A few others were seriously injured. Some of these people were shot by store-keepers protecting their property. Because neither the police nor the looters used guns, a disastrous night didn't turn into a fatally tragic one.

There were many minor injuries, though. While the police refrained from shooting anyone, they did use their clubs frequently. On occasion, the looters fought back with stones and bottles. Hospitals soon filled with the injured. But, however bloody the night was, it could have been worse. And it would have been worse if the looters had resisted arrest forcefully. On the whole, looters considered the police part of the "fun." They thought the night was something of a game between themselves and the authorities, a game of catch-us-if-you-can. Since they saw it as a game, they saw no reason to fight and get hurt. When they were caught, the majority of rioters went quietly to jail.

Most of the savagery that night was directed against property, not people, although at times it endangered lives, as well. The threat came from those rioters not satisfied with looting and vandalizing. They resorted to arson and burned everything from trash to whole buildings. In some places, the arson was bad enough to disturb other looters. Said one after it was all over, "When they got to the point where they started burning things down . . . that's when I started feeling: This is carrying a good thing too far."

Arson led to many injuries. Forty-four firefighters were hurt trying to bring blazes under control. Neighborhood residents also were injured by the fires, but many people were simply frightened and shocked. A few people packed up and left. A Bushwick man, for example, got his family together and took them to find a safer place to live. "We'll never come back here," his wife vowed. "Not even for a visit."

Dawn came, and the sun revealed to everyone the extent of the destruction. Though the night was over, the blackout and the violence it had wrought were not.

13

"This Creepy Feeling"

AFTER THE 1965 BLACKOUT, MAYOR ROBERT WAGNER didn't expect Con Edison to end power failures. But he did want the company to improve its system so that the next time a blackout struck it wouldn't last the whole night.

Twelve years later, another mayor and the rest of the city saw just how much progress had been made. The blackout didn't last the entire night; it went on much, much longer. At 9:00 A.M. Thursday morning, almost twelve hours after the lights went out, the city still lay under the shadow of the blackout. Mayor Beame examined the situation and declared New York City "closed for the day." Nothing like it had happened before. Snowstorms, at times, had paralyzed city streets, slowed the pace of business, and closed schools. But it took a power failure to shut down the entire city. The blackout closed subways, stock exchanges, theaters, and most stores and restaurants. Except for police officers, firefighters, and private guards, few people worked that day. Office buildings that never before had closed were shut tight. The financial district in the south of Manhattan, normally teeming with

Normally busy financial district on Wall Street, practically deserted on July 14   Wide World Photos

people on a business day, was eerily still. It looked like a scene from a movie about the end of the world.

No one knew when the blackout would end. Not the government. Not Con Edison. That fact was more disturbing than any other. "What bugs me," a visitor to the city told a reporter, "is that no one knows what's happening. It just keeps going on and on. I'm starting to get this creepy feeling."

A creepy feeling. Throughout the city, people felt it. It wasn't just because whole sections of the city were still being destroyed. It was also that the blackout was going on too long. One night, okay, but on and on it lingered, and for all people knew it might go on for days. As the thought sunk in, people became upset, angry, and afraid.

The blackout made normal life impossible. It stalled all air conditioners, and that was bad news on that hot Thursday. But by afternoon, people faced more serious problems. Many found themselves without two basic necessities of life: water and food.

The water stopped running because pumps that carried it to the top floors of city apartment buildings ran on electricity. Most buildings kept emergency water supplies in roof tanks. But the failure went on so long, the tanks soon emptied and people had nothing left. In fashionable neighborhoods, wealthy and powerful New Yorkers came out of their homes and joined poorer people in the search for water. Some found it; others had to do without it.

With refrigerators and freezers useless from the moment the power failed, food began to spoil. Everything that needed to be kept cold—meat, milk, frozen foods—turned to garbage in the heat of the day. Grocery stores lost enormous amounts of food. Restaurants, too, emptied their freezers and refrigerators, and tossed the contents into the nearest trash bin. While these businesses lost money, all New Yorkers faced the same problem. Much of the food they had on hand when the lights went out was no longer fit to eat on Thursday, and they had to find more.

Where? Many supermarkets didn't open at all on Thursday. Others had been looted and had little left to sell. Some did open, but because the owners feared robbers, they didn't

Street in Bedford-Stuyvesant section of Brooklyn covered with wreckage and debris the day after Wide World Photos

let customers inside. Instead, customers lined up outside the door. Store employees went out, took orders, and carried the food out. Though the day was wickedly hot, lines of hungry New Yorkers formed around the open stores. The people looked like refugees from a war.

A handful of restaurants opened, but those that did had little to offer. At one, the menu consisted solely of cantaloupe and coffee. By afternoon, many people were seen sitting in cafés where nothing at all was being served. They looked a little lost and confused. Without electricity for so long, people seemed to be losing their bearings. And so they sat and waited helplessly for normal—electric—life to return.

There were other signs that people were troubled and confused without their electric lifeline. For example, though the sun shone with painful brilliance, people strolled through the city's parks with flashlights in their hands. They clung to them as though they offered comfort and protection. Perhaps they regarded them as proof that human beings hadn't lost every means of creating light.

On Thursday, New Yorkers, knowing that the violence was still going on, worried that it would spread if the lights didn't return soon. If it reached them, people were ready to fight back. In areas bordering those already torn by riots, people patrolled the streets with iron pipes in their hands. At banks and large stores, guards prepared for attacks. Violence would be met by violence on Thursday night. "We're ready," said a security guard at one office building. The city seemed on the brink of a bloody battle.

Fears mounted during the day, as the rampage went on. Looters continued to roam the streets and tear through stores with glee. When television cameras caught them stealing, they didn't seem to care. Some even posed for the cameras, loot in hand. They still seemed to feel that what they were doing was perfectly all right.

As the violence went on, community leaders came forward to condemn it. Black and Hispanic leaders unanimously denounced the looters and the spirit of violence. Soon people added their voices to those of their leaders. They saw what this riot could mean to them. They knew that many ruined stores were never going to reopen. And if a neighborhood lost stores, it would lose jobs and money, as well. Those thoughts began to filter down to the streets and make a difference. Though the violence didn't end at once, it began slowly to wind down.

But what of Con Edison? Why was the company taking so long to bring back the electricity? Actually, a few thousand customers in Westchester had their power back within four hours after the start of the blackout. Getting it to the eight million people in the city took many more hours.

In New York City, the process of restoring power went slowly. In part, it was due to the size of the New York City system. Switches and cables, most of them underground, had to be inspected before Con Edison could reactivate them. And turbines had to cool and be brought back slowly. But Con Edison made mistakes that delayed the process further. In one case, engineers tried to link the Jamaica section of Queens with facilities on Staten Island. They botched the effort and, as a result, they again blacked out the entire borough of Queens.

Still, the utility made progress—slow, slow progress. By evening on Thursday, much of the city had power again. Some sections, however, remained in the dark. The blackout went into a second night.

One of the areas still blacked out was the East Side of Manhattan. It included Gracie Mansion, the residence of the mayor of New York. Asked by one mayor never to let a failure paralyze the city for twelve hours, Con Edison had kept another one in the dark for more than twenty-four hours. But, mercifully, the end was near.

14

"Christmas Is Over"

THE BLACKOUT FINALLY ENDED AT 10:39 P.M. ON Thursday, twenty-five hours after it had started. At that moment, New York City reopened and life returned to normal.

In the poor neighborhoods, the streets were quiet. The looting didn't spread, as had been feared. Instead, the violence died out. Most of the looters returned to their homes and resumed their everyday lives. As one young man said, "Christmas is over."

It had been an expensive Christmas for New York City. On Thursday and Friday, storekeepers saw the damage looters had done to their shops. More than 1,600 had suffered. Several shops had been reduced to rubble and could never reopen. Many merchants didn't plan on opening again even if they could. That was especially true of those who had stores in the worst-hit sections of the Bronx and Brooklyn. Some owners decided those neighborhoods had been ruined beyond repair. "You think anyone in his right mind would want to get back to this neighborhood?" one of them said. For him and others, the night of violence had been too awful to forget or forgive.

The post-blackout cleanup
Wide World Photos

In many parts of the city, a surprisingly large number of store owners announced that they would reopen. Some felt they had no choice. They had invested all their money in their stores, and they had none left with which to go elsewhere. A number of store owners simply decided that no rioters would drive them out of business. They returned to their stores on Friday and began cleaning up the mess. Soon after, the federal government stepped in to help them. The government officially declared New York City a disaster area. That meant that shopkeepers could get loans to help them reopen. More than a thousand shopkeepers filed loan applications.

Even with government money, storekeepers had a tough time getting back into business. Many had been hit so badly that repairs took months. One clothing store on the upper West Side, for example, was closed for eight months after the blackout.

A week after the blackout, the city tried to total the cost of the disaster. It proved hard to do. Rioters had done around fifty million dollars of damage. Other costs added perhaps one hundred fifty million dollars more to the total. But the long-term cost to the city was even higher. Some businesspeople, though unhurt by the riots, were scared by them. As a result they packed up and left town, taking their tax money and rent money out of New York for good. There were also individuals who left town to find work in what they saw as safer places, taking their money with them. Finally, there was the cost to the city of keeping hundreds of people behind bars to pay for their crimes of July 13 and 14.

On Thursday morning, July 14, New York City's jails overflowed with prisoners, and they stayed filled for several days. There were so many people in jail that the courts simply couldn't process them all. Usually, a prisoner is arrested and arraigned (formally charged with a crime) within a day. At the time of arraignment, most prisoners are permitted to leave jail on bail until their cases come to trial. That way, if they're found innocent later, they won't have had to spend more than a day behind bars. On July 13 and 14, the police arrested 3,700 people, and 3,700 was far too big a number to get through the courts quickly. Many prisoners had to wait days before being arraigned.

Meanwhile, conditions inside the jails were horrid. Prisoners were packed ten or more to a cell meant to hold four. Often they had to sleep on the floors. What's more, the hot weather turned the little cells into ovens. Temperatures soared well over 100 degrees every day.

There was, however, little sympathy for those in jail. New Yorkers knew and understood that the poor had spent a difficult summer. But people could understand the reasons for an act without approving it. And most New Yorkers did not approve of the blackout riots. They demanded that anyone found guilty of looting be severely punished. The *Amsterdam News*, the newspaper of New York's black community, called on the courts to sentence the guilty to "a year of hard labor," rebuilding the streets they'd destroyed. That creative suggestion didn't get much support. Most people wanted to see looters sent to prison for long periods of time. They wanted people to think twice about taking advantage of the darkness again.

As it turned out, judges pronounced tough prison sentences on some of the looters. Of the 3,700, a handful were convicted on serious charges. Most were let off entirely because there wasn't enough evidence against them. Others were convicted of minor crimes, and they received short jail sentences, fines, or probation. Bronx district attorney Mario Merola explained why the courts didn't throw more people behind bars: "The majority of people arrested were scavengers. The really bad guys who tore down the gates and ripped off the expensive stuff were gone when the police got there."

In the days after the blackout, New Yorkers felt anger at something else besides the looters: Con Edison. The utility tried to clear itself of blame for the blackout. Its chairman, Charles Luce, said that the outage had been "an act of God" and could not have been prevented. The public wasn't convinced. They generally agreed with Mayor Beame, who put most of the blame on the power company. Time and again, the mayor erupted in anger at Con Ed. At one emotional point, he even suggested that Charles Luce be hung. Throughout the blackout and for days afterward, the mayor continued to attack the company.

As usual, government officials ordered investigations to

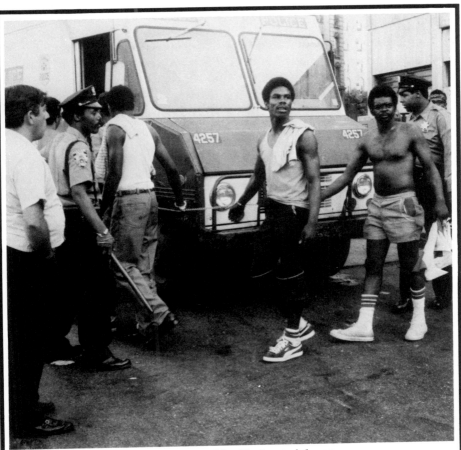
A few of the 3,700 people arrested for blackout violence
Wide World Photos

determine why and how it had happened. President Jimmy Carter called the blackout and the looting "intolerable," and he ordered the Federal Power Commission to investigate. The report the FPC finally issued did not deny that lightning had started the blackout. At the same time, the agency cast much of the blame on Con Edison. Investigators faulted the company's handling of the emergency and even questioned the ability of the operators at Con Ed's control center. A lot of people felt the report supported what they had thought from the beginning, that Con Edison again had let New York's electric system collapse.

The FPC report also contained ten suggestions to improve the system. The commission recommended more and better interconnections, better protective equipment, better emergency procedures, and the construction of new generators closer to the city's center. Con Edison seemed willing to make improvements, but New Yorkers doubted these would help much. They believed that the darkness would come again, probably soon. And they felt that the next time it came, things would likely get worse.

A new generator plant, powered by fuel cells
Consolidated Edison

Part IV

THE
FUTURE

15

Will the Darkness Come Again?

THE BLACKOUT OF 1977 HAD AN IMPACT ON Americans everywhere. The grim sight of ruined city streets shocked the country. People wondered if terror would rule every time the lights went out. The possibility seemed as real as it was horrifying.

Power companies hastened to reassure their customers. It couldn't happen there, companies outside New York insisted. Or if it did it wouldn't be as bad. It wouldn't last as long. There was some truth to this claim. New York City was so big, had so many people and so many miles of wires and cables, that it did take longer to bring back than any other system. But a mob of looters wouldn't need twenty-five hours. A few hours of darkness would be enough, and there wasn't a system in America that could promise it would never be out that long. Only days before the New York disaster, downtown St. Louis was out for eight hours. That blackout brought no violence, but it certainly could have. Eight hours was more than enough time for the violent to act.

In the few years since 1977, violence has not accompanied blackouts in the United States. There have been black-

131

outs, of course, but as of this writing none has involved looting, arson, or widespread vandalism.

During a recent blackout of the island of Puerto Rico, there were early reports of violence, although most proved untrue. At 8:15 P.M. on Friday, April 12, 1980, electricity throughout the island failed, and 3.5 million inhabitants lost power for as long as eighteen hours. At first, Puerto Rican officials believed that violence had *caused* the failure. Shortly after the lights went out, the governor of the island announced that political terrorists had apparently sabotaged the electric system. It seemed very likely because just before the blackout, terrorists had kidnaped an official of the Puerto Rican power company. Naturally, people assumed the two events were connected. They weren't. Once again, mechanical problems had knocked out an entire power system.

There was some violence during the power failure. Altogether, though, it amounted only to a few scattered incidents of vandalism and looting. The blackout was nothing like New York's in 1977. For the most part, the people of the island remained calm and quiet. Police officials in San Juan, the island's largest city, reported that the blackout produced the usual amount of chaos and confusion, but for the most part it was peaceful. Crime, they said, was "certainly no worse" than average for a Friday night. And no injuries were blamed on blackout violence.

After the power failure in Puerto Rico, the New York City blackout of '77 began to appear unusual. Clearly violence would not accompany every blackout or even most blackouts. Still, no one could say what would happen the next time a city or a region sank into the shadows of night.

And there surely will be a next time. The experts have stopped saying it won't happen again, and since March 29, 1979, it's doubtful anyone would believe them if they did.

On that day, the nuclear accident that environmentalists feared happened at a place called Three Mile Island in Pennsylvania. Some radioactive steam shot into the air after a valve on a power plant nuclear reactor had broken. The steam appeared to threaten the lives of everyone in the area, and people were evacuated as radioactivity spread over twenty miles. The

132

radioactivity was not bad enough to cause injury or disease, but people were justly frightened. And the reactor building, the last line of defense against a massive release of radioactive steam, was so filled with radioactivity that it had to be sealed up completely. No one opened the door for more than a year. It was an accident that many experts had confidently predicted would never happen.

Now that the experts and the utilities realize the system is not blackout-proof, they have redoubled their efforts to improve it. Con Edison, in particular, has labored to protect New York City from another disaster. Since 1977, it has made four clear improvements. First, the company has opened another high-voltage line to bring more power to the city from the P-J-M pool. Second, it has made its control center more automatic. Load-shedding is now being handled by equipment that should be faster and more reliable than the equipment that failed on the night of July 13, 1977, and Con Edison operators have new instructions to shed load quickly and decisively whenever a blackout looms. Third, the company has added quick-starting emergency generators to help restore the system rapidly in case of a blackout. And, finally, Con Edison has begun a close watch of any lightning activity in the New York area. When it starts, the company increases its spinning reserve to make sure that a stray bolt won't throw nine million people into darkness.

Con Edison and other power companies are also experimenting with new ideas and new machines that they hope will prevent or control power failures. One of these is to use direct current (DC) instead of alternating current (AC) for some long-distance transmission. Companies would still generate AC, because AC has many practical advantages over DC, and modern electrical fixtures and appliances can run on AC only.

We saw earlier that AC is easier to transmit, and it is. But DC is far stronger over long distances. Four to six times more electricity can flow through a DC line than through an AC line of the same voltage. It makes sense, then, for companies to generate AC current, convert it to DC for long-distance transmission, and convert it back to AC so people can use it.

With high-voltage DC lines, more power can flow to cities in times of need. An area in the Midwest that has power to spare can feed great amounts of it to a city like New York hundreds of miles away. What's more, if the New York Power Pool, or even the whole Northeast grid, began to experience a cascade again, the city would get power uninterruptedly through its long-line DC link to a distant source.

Why hasn't this been tried before? It has. A few DC transmission lines have been in operation for some time. But until recently, only one—a DC link between Washington State and Los Angeles—reached a major American city. This is because a facility to convert AC to DC and back to AC had to be immense. The reason is that electric current doesn't simply flow through conductors. It also jumps by means of an *arc* (a long, fiery spark) from one conductor to another, if the two are close enough together. The more powerful the voltage, the farther it will be able to jump. And if a piece of electrical equipment is struck by an arc from another, it can be severely damaged.

Insulation can prevent *arcing*. In our homes, wires are coated with materials like rubber and plastic to prevent arcing, though the power in any wall socket is so low that an arc couldn't go very far, anyway. High voltage equipment, on the other hand, carries tremendous amounts of power and sparks can jump from it and travel considerable distance to another conductor. (A conductor, you will recall, can be any piece of metal.) Yet high voltage equipment generally isn't coated because weather and time can cause a break in insulating materials. An arc, of course, could occur through the break.

There is, however, one reliable insulator for high voltage equipment—air. But air is pretty thin and to protect high voltage equipment you need a lot of it. As a result, AC/DC conversion stations require that their many pieces of equipment be placed far from one another. To operate safely, stations sometimes need as many as thirty acres of land, an area the size of more than twenty-five football fields! That much space might be available in rural areas, but it isn't available in crowded cities like New York.

Lately, however, scientists have been experimenting with

new insulators to protect conversion equipment. One that shows particular promise is a gas—like the air—called *sulphur hexafloride*. The gas prevents arcing when equipment is put in containers filled with it. Using this gas, a power company can bring pieces of equipment near one another. Con Edison has been building a conversion station in Queens that will use sulphur hexafloride containers. The facility will pick up long-distance DC transmissions and convert the power back to AC for use in New York. The station will measure only 60 feet by 130 feet—smaller than many factory buildings.

Another hope for the prevention and control of blackouts is a new type of power station. Electricity in these stations won't be produced by spinning steam turbines or water power. Instead power will be produced by giant *fuel cells.* Fuel cells are like huge batteries that turn chemical energy into electrical energy. Typically, a battery consists of two metal plates—called *electrodes*—stuck into a chemical solution. The solution reacts with the metals, filling one with positive charge and the other with negative charge. The two charged plates cannot discharge through the chemical solution. If a wire is touched to an end of both, however, the electrons in the negative electrode flow toward the positive, creating a continuous current. The chemical solution continues to charge the electrodes for a certain period of time. But, after a while, the chemicals undergo changes themselves, and they lose the ability to charge the electrodes. We say then that the battery is dead.

Fuel cells work essentially the same way, but with one important difference. They can't die because the chemicals inside them are changed constantly. Think of an ordinary flashlight battery: The chemical solution inside it is in the form of a paste. When it stops reacting, the battery becomes useless. But if the paste could be changed every day, the battery would never run down. That is how fuel cells operate. They never die because they get a continuous supply of electricity-producing chemicals. Old chemicals are simply flushed out and thrown away.

Fuel cells can help prevent blackouts because they are quick starting. Ordinary steam generators, as we've noted, take time to start. Hours. Attempts to bring them along faster can damage them. Keeping a large number of generators going as spinning reserve, though possible, costs a lot of money. As a result, spinning reserve is always limited and often isn't enough when a major crisis occurs. Fuel cells, on the other hand, can come up to full power in a matter of seconds. As soon as a blackout looms, a utility can turn to fuel cells and get instant results.

Since fuel cells have other advantages (for example, they don't cause air pollution) you might wonder why power companies don't produce all their electricity by fuel cells. The reason is that fuel cells are weak. The power they supply would be a drop in the bucket for any city. A new fuel cell plant under construction in New York City, for instance, will produce about forty-five hundred kilowatts. A conventional generator like Big Allis can churn out as much as one million kilowatts. Even a thousand fuel cell plants couldn't produce enough power for New York City. And Con Edison cannot afford—and New York City doesn't have room for—one thousand plants. As engineers improve the cells, they might be able to produce more power, though probably not enough to generate all of a city's electricity. However, a few fuel cell stations might help prevent blackouts or at least shorten them.

Power companies also have been searching for ways to "store" electricity. Generators often sit idle. Ideally, power companies would like to be able to turn them on during slow hours, store the extra electricity, and release it during peak load or emergencies. As we learned, electric power can't be stored, not really. It has to be made the moment it's needed.

There are, however, ways to store electricity indirectly. One method is quite common. Many household gadgets, like pocket calculators, have what are called *rechargeable* or *storage* batteries. These batteries work like other batteries; they produce electricity through chemical changes. But when they begin to run down, they can be restored by feeding electric current into them. The electricity isn't captured and tucked away in a corner. Rather, it alters the chemicals inside the battery so that they can charge the electrodes and produce current again.

Power companies can feed electricity into storage batteries during hours when they have power to spare, say after midnight. During a time of need, the batteries can then be switched on to provide extra electricity. When the need is satisfied, they can of course be recharged. But storage batteries, like fuel cells, produce little electric power at present. One test system in New Jersey produces two thousand kilowatts with banks of batteries. Still, a few such facilities might come in handy.

There is another method of storage that can provide large amounts of extra power. Called *pumped storage*, it would be the ideal means of storing electricity, except that pumped storage facilities require huge amounts of money, land, and water, and building them can damage the environment. Pumped storage works this way: During hours when a power company has extra electricity, it uses some of that power to pump water to high ground. Then, when the company needs power, it allows the water to flow downhill and has it turn a generator as it flows. A pumped storage facility can add a million or more kilowatts to a system for a short period of time. However, pumped storage really requires the building of an artificial lake, and building a lake is no simple matter. Also, it means taking great amounts of water out of one place and putting the water someplace else. Often that can damage the environment. In the 1960s, environmentalists prevented Con Edison from building a pumped storage facility in New York State.

Still, all of these ideas and devices may help utilities deal with sudden power emergencies. They may not prevent blackouts entirely, but perhaps they will help keep them short and localized. If they can do either, they'll be important.

Sudden power emergencies, however, are only one aspect of the story of blackouts. Shortages also produce blackouts, and further shortages are possible. Remember, the threats to our supplies of electricity are twofold: Power companies might not be able to get enough oil and possibly natural gas, and they might not be able to build enough plants to meet demand. Solutions to both seem obvious. The electric industry should use less oil, and people should demand less power. Simple? Unfortunately, too simple.

Power companies want to get away from a dependence on

oil and gas. They would like to switch oil-burning plants to coal, a fuel America has in abundance. And they would like to build only coal, hydro, and nuclear power plants in the future. But coal, of course, produces dirty smoke. And while the smoke can be cleaned sufficiently, the equipment that does it is expensive. Also, to convert an oil-burning plant to a coal-burning plant would cost power companies a lot of money. Utilities might make the changeover if the government helped pay the cost or at least loaned them some of the money. But right now, because of pollution, some government officials aren't sure that the companies should convert at all. Certainly, neither the government nor the utilities are ready to spend a great deal of money on something that might prove a bad idea in the long run.

As for future building projects, coal, water, and nuclear plants may have certain advantages over those needing oil or gas. But coal and nuclear power, as we've seen, have disadvantages. Water power, too, has some drawbacks. It would be fine if there were many available waterfalls and rushing rivers, but there aren't. Some rivers can be dammed and man-made waterfalls created. But often dams cause harm to the environment. One dam project in Tennessee, for example, was held up for years because when completed it could have killed off a kind of fish called a snail darter.

Power companies are experimenting with new and different types of fuels, and these hold promise for the future. In some parts of the country, there are natural hot-water springs. They're hot enough to create steam to run turbines. Other parts of the nation have steady winds, and they, too, can turn turbines. Also, there are ways that sunlight can either create heat for steam or be turned directly into electricity. These methods are clean and have no known dangers. But they are still experimental and probably won't make a difference in our electricity supplies for some time to come.

At present, coal is probably the only fuel we have that offers a real alternative to foreign oil. Natural gas might also prove more plentiful than people used to believe, but that remains to be seen. Coal smoke, if not thoroughly cleaned, may be dangerously dirty, but if Americans keep using more and more electricity, power companies will have little choice but to build more coal plants. They may build possibly unsafe nuclear plants, as well.

If Americans use less electricity, however, the electric industry won't have to build more plants. It won't have to use more oil. Utilities will be able to convert oil-burning plants slowly to coal. In the meantime, they will be able to improve smoke cleaners so that the coal won't seriously pollute our air. And they will be able to produce cleaner alternatives, too. In ten or twenty years, perhaps, oil and coal plants will be replaced by those that use the power of the sun and wind. No more nuclear plants will need to be built. We won't have to expose any more people to the dangers of a nuclear accident.

Unfortunately, to Americans the idea of using less electricity is not appealing. Most of us grew up in a world in which we could use more and more all the time. We could waste power. We could run air conditioners hard, so that the temperature would drop below 70 degrees Fahrenheit. We could let lights burn for hours, whether or not anyone was using them. And it's very hard to change our habits.

Of course, individuals aren't the only ones who waste electricity. Businesses do, too. Stores let banks of lights and televisions go all night long to attract attention. And some factories have poor machinery that uses more electricity than it needs.

But there are signs that both businesses and individuals realize they can't go on wasting power. Changing habits might be difficult and even painful, but a lot of Americans appear willing to do it. In the first months of 1980, use of electricity actually dropped. And while an unusually hot summer brought usage back up, it's clear that people are thinking about saving power. Factories are installing new power-saving equipment; stores are turning off lights. People generally are trying to remember that every time they waste power, they, and maybe the entire country, pay for it.

Actually, there are fears that if we use less, the country will become poorer as a result. The electric companies convinced Americans many years ago that if we wanted to grow, we had to use more electricity. Some people still believe it.

Less use of power, they say, means that people are buying less, making less, doing less. But that isn't necessarily true. Other countries, like Japan, have had to conserve for years. Having no fuels of their own to burn, they've had to spend lots of money buying oil and coal from other countries. Energy has always been expensive in Japan. Yet the Japanese have prospered. In recent years their industry has grown faster than ours. We have always wasted tremendous amounts of electricity, and we do not have to waste to grow. We can use less electricity more efficiently. If we do that, we might be taking a step toward preventing power shortages in the future. Shortages will only make our country poorer. But if we learn to make do with less, we will continue to grow and prosper.

No matter what we do, or what the utilities do, however, we probably will never see the end of power failures. Maybe they'll come less often, but they'll come. Again, systems break down. Equipment fails. People make mistakes. Electric power, after all, is something controlled by humans. When it goes, though, we should not feel helpless. The switch on the wall may determine what we do; it shouldn't determine how we feel.

In the darkness, strange places can terrify even the bravest souls momentarily. But no matter where a person is during a blackout, he or she needs only calm and intelligence to make it through. Experience has shown that people can survive quite easily even in pitch-dark subway tunnels or cramped, stalled elevators.

Violence *is* something to fear. There was a special set of circumstances that led to the riots of July 13, 1977. Maybe riots won't accompany blackouts again. But the only way to ensure this is to make sure the circumstances are changed. Remember that the rioters didn't care about their neighborhoods. They didn't feel they belonged to the streets they tore apart. In neighborhoods where people felt a part of a community, there was no violence. Perhaps if more people are given jobs and secure homes, violence will not erupt again in a blackout.

Probably, most of us will go through our next blackout without seeing any signs of the "stick-up man." We will be a little surprised that the switch on the wall has failed us; we always are. But we should not find any cause for fear. Only the darkness will have come, and the darkness itself is nothing to be afraid of.

INDEX